Biology, Epidemiology and Management of *Pyrodinium* Red Tides

Proceedings of the Management and Training Workshop
Bandar Seri Begawan, Brunei Darussalam
23-30 May 1989

Edited by

Gustaaf M. Hallegraeff
and
J.L. Maclean

1989

FISHERIES DEPARTMENT, MINISTRY OF DEVELOPMENT
BANDAR SERI BEGAWAN, BRUNEI DARUSSALAM

INTERNATIONAL CENTER FOR LIVING AQUATIC RESOURCES MANAGEMENT
MANILA, PHILIPPINES

Biology, epidemiology and management of *Pyrodinium* red tides

Proceedings of the Management and Training Workshop
Bandar Seri Begawan, Brunei Darussalam
23-30 May 1989

Edited by

G.M. HALLEGRAEFF
and
J.L. MACLEAN

1989

Printed in Manila, Philippines

Published by the Fisheries Department, Ministry of Development, P.O. Box 2161, Bandar Seri Begawan 1921, Brunei Darussalam, and International Center for Living Aquatic Resources Management, MC P.O. Box 1501, Makati, Metro Manila, Philippines

Hallegraeff, G.M. and J.L. Maclean, Editors. 1989. Biology, epidemiology and management of *Pyrodinium* red tides. ICLARM Conf. Proc. 21, 286 p.

ISSN 0115-4435
ISBN 971-1022-64-8

Cover: Electron micrograph of *Pyrodinium bahamense bahamense*, Oyster Bay, Jamaica, from a sample collected by R.J. Buchanan on 17 March 1967. Photograph by Gustaaf Hallegraeff.

ICLARM Contribution No. 585.

Biology, Epidemiology and Management of *Pyrodinium* Red Tides

Contents

Research, Monitoring and Equipment

Management

Epidemiology

Economics

Part 2. Manual for Field and Laboratory Research on the Dinoflagellate Pyrodinium bahamense

Identification

Field Methodology

Laboratory Techniques

Epidemiology

Foreword

Red tide has become a frequent phenomenon in the coastal waters in many parts of the world. In Southeast Asia many lives have been lost and thousands taken ill from eating contaminated seafoods. Economic losses are also serious and effective mitigating measures are greatly needed.

In 1984 two regional meetings, a WESTPAC Red Tide workshop at CSIRO in Cronulla (Australia) and a SEAFDEC/IDRC meeting on Toxic Red Tides and Shellfish Toxicity in Southeast Asia held in Singapore, unanimously declared *Pyrodinium bahamense* as the "number one" red tide danger in the Indo-West Pacific region. Urged by the 1987 Guatemala poisonings and the first appearance in 1988 of red tides in Manila Bay (Philippines), the concept of a meeting devoted exclusively to *Pyrodinium bahamense* was first proposed by Dr. Gustaaf M. Hallegraeff of CSIRO and Mr. J.L. Maclean of ICLARM. The ASEAN/US Coastal Area Management Project took up the proposal and developed it into a *Pyrodinium* red tide management and training workshop which permitted the participation of red tide researchers, public health officers and administrators to discuss research and management issues on *Pyrodinium* red tides and to develop a training manual.

Brunei Darussalam readily hosted the workshop and contributed to part of local costs of the workshop. The idea found ready support from a number of donor agencies: The Australian International Development Assistance Bureau (AIDAB), the United Nations Environment Program (UNEP), Intergovernmental Oceanographic Commission (IOC/WESTPAC), the National Academy of Sciences, USA, International Development Research Centre (IDRC) of Canada and the United States Agency for International Development (USAID). As a result, from 23 to 30 May 1989, over 40 researchers from the six Asean countries (Brunei Darussalam, Indonesia, Malaysia, the Philippines, Singapore, Thailand), and from Australia, Canada, Japan, Papua New Guinea, Central America and the USA gathered in Brunei Darussalam to discuss the biological, economic, management, medical and training issues of *Pyrodinium* red tides.

We were pleased that the workshop was successfully concluded and the proceedings published within a period of six months. Credit should be given to the enthusiasm and untiring efforts of Dr. Hallegraeff and Mr. Maclean as well as technical support staff of the Department of Fisheries, University of Brunei Darussalam and ICLARM.

Haji Matdanan bin Jaafar
Director
Department of Fisheries
Brunei Darussalam

Chua Thia Eng
Project Coordinator
ASEAN/US Coastal Resources
Management Project, and Director
Coastal Area Management
Program, ICLARM

Preface

The dinoflagellate *Pyrodinium bahamense* Plate 1906 (*Pyro* = fire; *bahamense* = from the Bahamas) was first described from the tropical Atlantic more than 80 years ago. In Bahia Fosforescente in Puerto Rico and Oyster Bay in Jamaica, this species forms persistent luminescent blooms which are a major tourist attraction. Residents in these areas eat small oysters attached to mangrove roots, apparently without ill effects. The first harmful implications of *Pyrodinium* blooms became evident in 1972 in Papua New Guinea. Red-brown water discolorations coincided with the fatal food poisoning of three children and mouse bioassays on shellfish from a house in the affected village subsequently established *Pyrodinium bahamense* as the source of paralytic shellfish poisons. Since then, toxic *Pyrodinium* blooms have apparently spread to Brunei and Sabah (1976), the central Philippines (1983) and the northern Philippines (1987). Most unexpectedly, in 1987 on the Pacific coast of Guatemala, 187 people had to be hospitalized after consumption of toxic clams and 26 persons died. The problem was initially attributed to pesticide poisoning and only later linked to *Pyrodinium.* Altogether, this species has now been responsible for more than 1,000 human illnesses and 60 fatalities resulting from the consumption of contaminated shellfish as well as planktivorous fish such as sardines and anchovies. Unfortunately, the tropical countries in the Indo-West Pacific (Brunei, Indonesia, Palau, Papua New Guinea, Philippines, Sabah, Solomon Islands) and Latin America (Guatemala, Venezuela) that are affected (Fig. 1) depend heavily on seafoods for protein and have little prior experience in toxic dinoflagellate research.

The present workshop was conceived to bring together the researchers, managers and medical practitioners who have been thrust into this little-understood area in order to consolidate knowledge on all aspects of *Pyrodinium* red tides and their effects and to discuss solutions to the problem.

Invited experts reviewed current knowledge on *Pyrodinium* taxonomy (Taylor, Fukuyo), its benthic resting cyst (Anderson, Matsuoka), biology (Seliger) and toxicology (Oshima, Hall). Round-table discussions compared regional management plans and reviewed

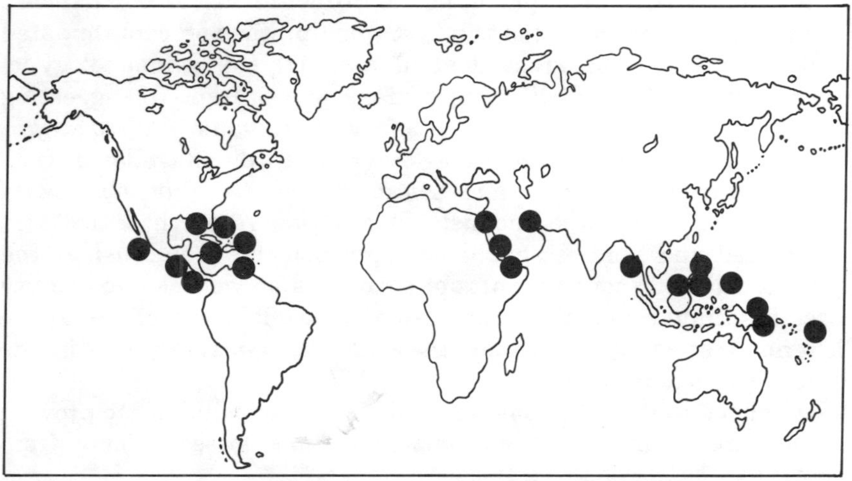

Fig. 1. Global distribution of *Pyrodinium bahamense* in recent plankton. Distribution of cysts is shown on p. 224.

epidemiology and economic effects. Following the three days of talks, training sessions were held in dinoflagellate and cyst taxonomy, dinoflagellate culturing, toxicology (mouse bioassay and HPLC) and the use of oceanographic field equipment.

All in all it was a week of cooperative brainstorming and research in which theories were developed and future experiments proposed. An important question still to be answered is whether populations from the tropical Atlantic (var. *bahamense*) are identical to those from the Indo-West Pacific (var. *compressum*) or whether there are analogies with the *Alexandrium tamarense/catenella* group which exhibits considerable genetic diversity resulting in variability in toxin and isozyme profiles, chain-formation, luminescent vs non-luminescent strains, etc. This can significantly confuse experimental and field observations. Howard Seliger, who carried out extensive studies on blooms in Caribbean locations, interviewed researchers present at the workshop and sifted through data and maps to formulate some general theories on *Pyrodinium* blooms in the Indo-West Pacific. Do *Pyrodinium* blooms originate inshore (e.g., in mangrove areas) or offshore and what is the role of benthic resting cysts in seeding red tides? During the workshop, field trips to Brunei Bay and neighboring Sabah waters (Kuala Penyu) discovered cysts but only in very low concentrations. A survey of fossil occurrences of this cyst (traceable to the Eocene, 50 million years ago) indicates a much wider range of distribution in the past than known at present.

Using the new HPLC equipment at the Brunei Fisheries Department, Yasukatsu Oshima produced the first toxin profiles on contaminated planktivorous fish. A great deal of data on the epidemiology of Philippines paralytic shellfish poisonings was presented, suggesting possible differences in the susceptibility of humans to *Pyrodinium* toxins compared to those from other PSP dinoflagellates (i.e., *Gymnodinium catenatum* and *Alexandrium* (= *Protogonyaulax)* species). The need for economists in red tide research teams was emphasized and studies should be commissioned that consider the actual social and economic costs of outbreaks as well as opportunity losses associated with the suppression of development of resources. Fisheries managers put up for inspection their action plans in the event of future outbreaks.

We hope that the lessons from this workshop will help to prevent loss of human lives and minimize economic losses arising from *Pyrodinium bahamense* red tides.

Gustaaf M. Hallegraeff
Jay L. Maclean
December 1989

Part 1

Proceedings of the Management and Training Workshop, Bandar Seri Begawan, Brunei Darussalam, 23 to 30 May 1989

An Overview of *Pyrodinium* Red Tides in the Western Pacific

J.L. MACLEAN
International Center for Living Aquatic Resources Management
MC P.O. Box 1501
Makati, Metro Manila
Philippines

MACLEAN, J.L. 1989. An overview of *Pyrodinium* red tides in the western Pacific, p. 1-8. *In* G.M. Hallegraeff and J.L. Maclean (eds.) Biology, epidemiology and management of *Pyrodinium* red tides. ICLARM Conference Proceedings 21, 286 p. Fisheries Department, Ministry of Development, Brunei Darussalam, and International Center for Living Aquatic Resources Management, Manila, Philippines.

Abstract

Red tides and paralytic shellfish poisoning caused by *Pyrodinium bahamense* var. *compressum* have been recorded in the tropical western Pacific since 1972, although it appears certain that there were previous outbreaks in Papua New Guinea (PNG) at least. The incidence of red tides seems to be associated with one monsoon or the other in each country. Wind-driven upwelling is proposed to be a dominant factor in bloom initiation. The sequential appearence of toxic *Pyrodinium* blooms in PNG, Western Borneo and the Philippines suggests a northerly spread of blooms if not of the organism itself. Coincidence of major blooms with ENSO event years suggests an association of the two phenomena and perhaps a northerly movement of ENSO anomalies since the 1970s.

Introduction

The following summary of *Pyrodinium* red tides and PSP is brief because most of the information has appeared in earlier publications which covered various time periods (Maclean 1979, 1984, 1985, 1989), while new information has been summarized from other contributions to this volume.

Distribution

Pyrodinium was unknown in the western Pacific until 1972 when blooms of *Pyrodinium bahamense* var. *compressum* were found to be responsible for paralytic shellfish poisoning (PSP) in the vicinity of Port Moresby, Papua New Guinea (PNG). A search of government records indicated that PSP had been present in PNG since at least 1927. Several potentially harmful bloom species occur in PNG waters (Appendix I in Maclean, this vol.), but only *P. bahamense* has been found in visible concentrations.

Pyrodinium red tides and/or PSP had been observed occasionally in other parts of PNG, generally all within the northwest monsoon season, November-June. No confirmed reports of *Pyrodinium* blooms have been recorded since 1974. The only conclusion to be drawn from this is that there have been no large outbreaks of PSP since 1972.

However, in January-May 1976, major toxic *Pyrodinium* red tides were found for the first time along about 300 km of the northwestern coast of Borneo, encompassing Brunei Darussalam and Sabah, Malaysia. PSP but no further red tides occurred in this area until 1980 when there were sightings associated with PSP in April-June and December 1980-January 1981.

In April-May 1981, toxic *Pyrodinium* blooms were discovered on the east coast of Borneo (Sabah). They have not been found there since then but there were PSP cases in adjacent Indonesian waters of eastern Borneo in 1988 (Q. Adnan, pers. comm.).

The western Borneo coast was free of red tides in 1982-1984; they were present again in 1985, 1986 and 1988. PSP, however, has been detected in the intervening years, suggesting that sub-visible concentrations of *Pyrodinium* are probably present every year.

Toxic *Pyrodinium* red tides were first found in the Philippines in mid-1983, in the Samar Sea. A second major outbreak occurred there in mid-1987. On this occasion *Pyrodinium* blooms were also detected in Zambales, some 500 km north. In August-October 1988, *Pyrodinium* red tides appeared in Manila Bay, just south of Zambales. They were reported in the Samar Sea by September and remained present in parts of the central Philippines until March 1989. All these events resulted in many illnesses and deaths.

Borneo, the Philippines and PNG are the major known *Pyrodinium* problem areas in the western Pacific. There have been PSP outbreaks in Indonesia apart from that in eastern Borneo. However, the organism involved has not been determined. In Palau, *Pyrodinium* may be continuously present in deep blooms in Arumizu Bay, while PSP or red tides of unknown origin have occurred in Fiji, Guam, Samoa, the Solomon Islands and Tuvalu.

A summary of these events is given in Fig. 1.

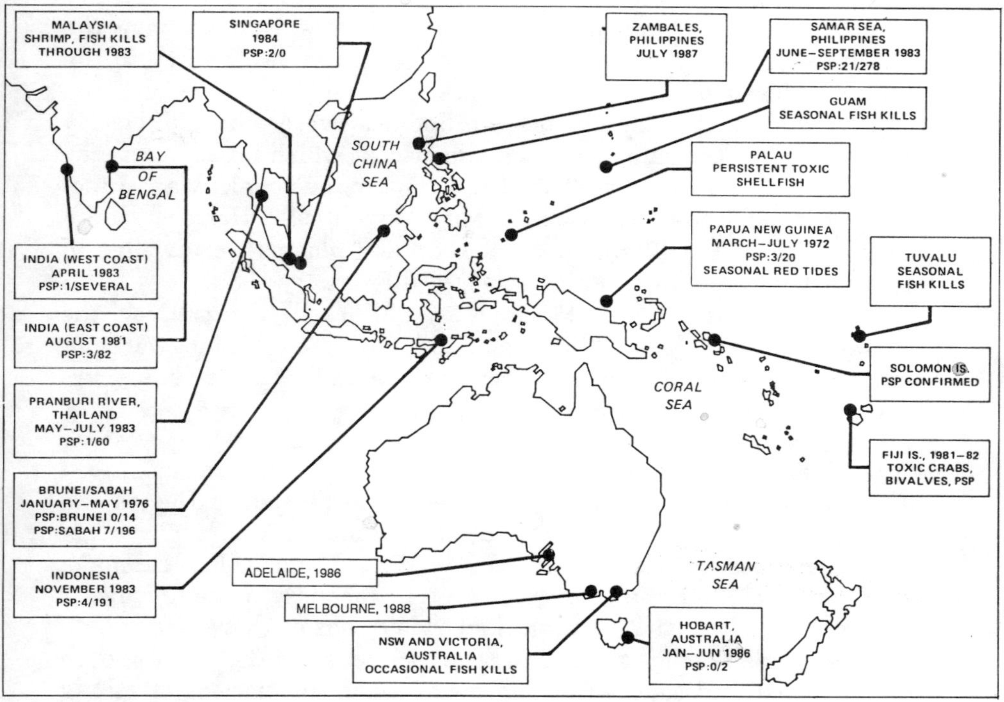

Fig. 1. Toxic Indo-Pacific red tides, fish kills and paralytic shellfish poisoning (PSP). Numbers and dates in boxes refer to number of deaths/illnesses and time of first reported incidents. (Updated from Maclean 1989).

Seasonality

In Port Moresby, blooms of *Pyrodinium* occurred seasonally in the rainy months (January-May) during four years of observations (1971-1974). In fact all observed or suspected blooms have taken place within the northwest monsoon period, which can begin as early as November and end as late as July.

Seasonality of *Pyrodinium* blooms in the Sabah-Brunei Darussalam area is more difficult to interpret. Toxic red tides on the northwest coast of Borneo have all occurred between December and June, corresponding to the northeast monsoon. However, PSP cases have been detected in most other months of the year. To complicate matters further, the *Pyrodinium* population in Kuala Penyu lagoon in southwest Sabah increased in the latter half of the three years of investigations (1986-1988) with major peaks in August-September. A much smaller peak was observed there in May 1986 and February 1988. Why the lagoon and the coastal waters appear to have the opposite seasonality remains a mystery that may be the key to the ecology of the species.

Pyrodinium blooms in the Philippines were initially clearly associated with the rainy southwest monsoon, which broadly extends from May to December. The first Samar outbreak was June-September 1983; the second was in May-August 1987. However, it did not begin there until September in 1988 and blooms were visible in other parts of the central Philippines until February 1989.

Further north in the Philippines, PSP cases occurred on the Zambales coast from April to August 1987, although blooms were not sighted until July. There had been PSP in the two previous years there also.

A common feature of all the sites of red tides in the western Pacific has been the presence of mangroves in the general vicinity. I know of only one exception, the western side of Manila Bay in the Philippines where, however, there used to be mangroves.

Other associations are tenuous or contradictory. For example, Port Moresby, PNG, red tides were present for most of the northwest monsoon, which is the very marked rainy season there. In Sabah, visible red tide patches seemed to be associated with the onset of both the northeast and southwest monsoons, which are periods of strong winds. Sabah red tides apparently need an intervening period of calm, sunny weather before blooming. Continuous sunny weather was associated with the decline of the 1983 Philippine blooms (Estudillo and Gonzales 1984). PNG and Philippine blooms in general follow the onset of rains. Rain inhibited Sabah red tides.

An offshore wind direction during the general red tide "season" characterizes the Port Moresby area and the west coast of Sabah. Conditions in general would be more sheltered and local inshore upwelling would be promoted. Philippine red tides have been associated generally with rainy weather and onshore winds. However, the prevailing winds in the first Samar red tides were offshore (Estudillo and Gonzales 1984).

One possible conclusion is that wind is the more important factor and when it begins to blow in the right direction with adequate strength, cysts and or nutrients from the substrate are brought up into the water column to initiate blooms. The occurrences of PSP in the "wrong" season in, e.g., Sabah would suggest that the *Pyrodinium* population does well all year round there and that wind-driven upwelling simply catalyzes the formation of red tides. On the other hand, where *Pyrodinium* disappears from the water column in the off-season, upwelling is a convenient mechanism to bring cysts and new nutrients together into the water column.

Habitat

Although neritic, *Pyrodinium* prefers more saline waters, 24.7-36.8 ppt in PNG, 31.1-34.9 ppt in the first Samar, Philippines, blooms (Estudillo and Gonzales 1984), and 24.3-32.1 ppt in Borneo (Beales 1976). Culture trials gave best performance in 100% seawater (Blackburn and Oshima, this vol.).

Water temperatures in the various red tide areas have ranged between 24.4 and 32.5°C, indicating the tropical nature of *Pyrodinium bahamense* var. *compressum*. Interestingly the *bahamense* variety is found in subtropical temperatures down to 22.2°C in Florida (Steidinger and Williams 1970) although in Jamaica it is found in much warmer water temperatures, 27.0-35.0°C (Buchanan 1971). In culture, the *compressum* variety grew well at 25°C and 30°C (Blackburn and Oshima, this vol.).

Movement

Descriptions of red tides in the three major study areas indicate variously that blooms form offshore and move or are blown inshore (Brunei Darussalam, Sabah); form inshore and drift offshore (Samar); or remain where they are formed (PNG). In Sabah, patches were observed on one occasion to drift with the surface current and remain intact. In the Philippines, patches were thought to move across the Samar Sea.

Vertical migration of *Pyrodinium* has been well established for red tides in Port Moresby, PNG, rising to the surface in mid-morning and sinking in mid-afternoon. Dinoflagellates in general and *Pyrodinium* in particular are known to use their vertical migration ability to avoid being flushed out of bays. Does the opposite effect take place to a significant extent?

The extent of horizontal movement of *Pyrodinium* (vegetative) cells is an important issue because the pattern of red tides in the western Pacific can be interpreted as a spreading of the organism's distribution from PNG (where it could have arrived in ballast of, e.g., copra vessels from the species' previously known habitat in the Red Sea area; ships would take on ballast there after passing eastward through the Suez Canal) to Borneo (1976) and to the Philippines (1983). The general oceanographic features of the area would allow this interpretation but not explain why there have not been outbreaks at intermediate locations. The blooms in Palau would require a different explanation. The strong equatorial current would provide a quick transport mechanism from Palau to the Philippines but what is the origin of Palau's *Pyrodinium?* On the other hand, dinoflagellates are known to survive long trips in ship's ballast water (Hallegraeff et al. 1988) and copra vessels may also be the vector in the case of Palau.

One problem in the "drift" hypothesis is the presence of fossil cysts over a wider geographic area from Australia to Israel in the Eocene period (see Discussion and Recommendations on Research, p. 00). Further, the western Pacific form of *Pyrodinium* tends to be flatter than elsewhere and is treated as a separate variety by some authors. Finally, *Pyrodinium* is a neritic rather than an oceanic species.

In a recent article (Maclean 1989), I pointed out the exact coincidence of the major toxic red tides between 1972 and 1987 with ENSO (El Niño-Southern Oscillation) events (Fig. 2). My interpretation was that ENSO events probably affect different parts of the western Pacific to varying degrees on each occasion, creating a short- or long-term environment suitable for toxic blooms in one locality or another.

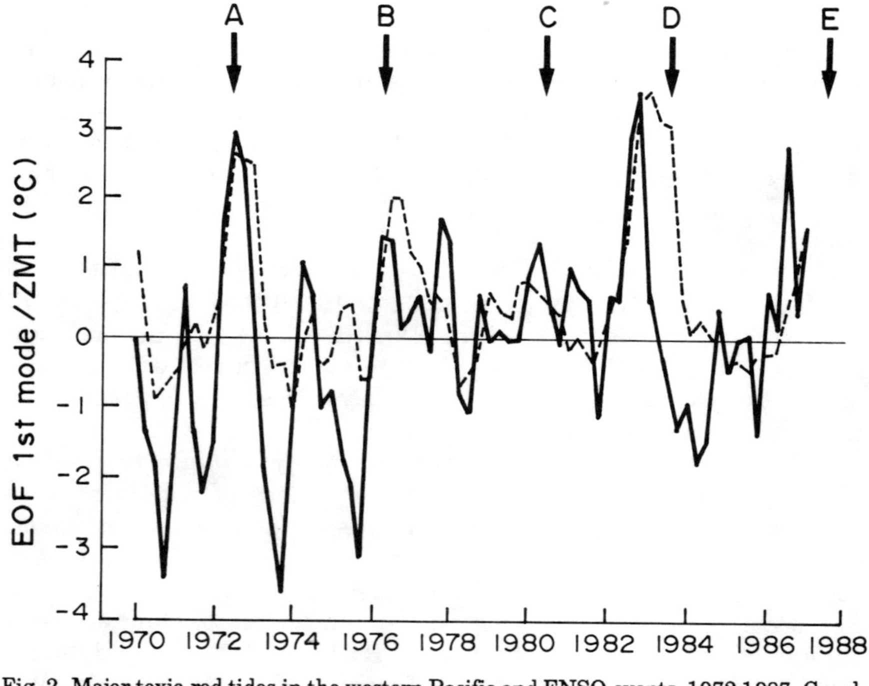

Fig. 2. Major toxic red tides in the western Pacific and ENSO events, 1972-1987. Graph shows empirical orthogonal functions (EOF, solid lines) of zonal wind anomalies, and zonal mean surface temperature anomalies (ZMT, broken line) over the near-equatorial eastern Indian and western Pacific Oceans. Strong positive anomalies are indicative of ENSO events and are seemingly correlated with *Pyrodinium* red tide events. Arrows show time of onset of red tides: A: Papua New Guinea, 1972; B: Borneo, 1976; C: Borneo, 1979-1980; D: Philippines, 1983; E: Philippines, 1987. (Modified from Maclean 1989)

It may be that these events in the western Pacific as they relate to oceanographic conditions have been moving northward. This would explain the apparent absence of red tides now in PNG.

References

Beales, R.W. 1976. A red tide in Brunei's coastal waters. Brunei Mus. J. 3(4): 167-182.

Buchanan, R.J. 1971. Studies at Oyster Bay in Jamaica, West Indies, V. Qualitative observations on the planktonic algae and protozoa. Bull. Mar. Sci. 21(4): 914-937.

Estudillo, R.A. and C.L. Gonzales. 1984. Red tides and paralytic shellfish poisoning in the Philippines, p. 52-79. *In* A.W. White; M. Anraku and K.-K. Hooi (eds.) Toxic red tides and shellfish toxicity in Southeast Asia. Southeast Asian Fisheries Development Center, Singapore, and International Development Research Centre, Ottawa, Canada.

Hallegraeff, G., C. Bolch, B. Koerbin and J. Bryan. 1988. Ballast water a danger to aquaculture. Aust. Fish. 47(7): 32-34.

Maclean, J.L. 1979. Indo-Pacific red tides, p. 173-178. *In* D.L. Taylor and H.H. Seliger (eds.) Toxic dinoflagellate blooms. Elsevier Science Publishing, New York.

Maclean, J.L. 1984. Indo-Pacific red tide occurrences, 1972-1984, p. 92-98. *In* A.W. White, M. Anraku and K.-K. Hooi (eds.) Toxic red tides and shellfish toxicity in Southeast Asia. Southeast Asian Fisheries Development Center, Singapore, and International Development Research Centre, Ottawa, Canada.

Maclean, J.L. 1985. Toxic dinoflagellate blooms in Asia, a growing concern, p. 517-520. *In* D.M. Anderson, A.W. White and D.G. Baden (eds.) Toxic dinoflagellates. Proceedings of the Third International Conference on Toxic Dinoflagellates, Brunswick, Canada, 8-12 June 1985. Elsevier Science Publishing, New York.

Maclean, J.L. 1989. Indo-Pacific red tide occurrences, 1985-1988. Mar. Pollut. Bull. 20(7): 304-310.

Steidinger, K.A. and J. Williams. 1970. Memoirs of the *Hourglass* cruises. Vol. 2. Marine Research Laboratory, Florida Department of Natural Resources, St. Petersburg, Florida.

Pyrodinium Red Tide Occurrences in Brunei Darussalam

MATDANAN HAJI JAAFAR
M.W.R.N. DE SILVA
P.H.Y. SHARIFUDDIN
Department of Fisheries
P.O. Box 2161
Bandar Seri Begawan 1921
Brunei Darussalam

MATDANAN HAJI JAAFAR, M.W.R.N. DE SILVA and P.H.Y. SHARIFUDDIN. 1989. *Pyrodinium* red tide occurrences in Brunei Darussalam, p. 9-17. *In* G.M. Hallegraeff and J.L. Maclean (eds.) Biology, epidemiology and management of *Pyrodinium* red tides. ICLARM Conference Proceedings 21, 286 p. Fisheries Department, Ministry of Development, Brunei Darussalam, and International Center for Living Aquatic Resources Management, Manila, Philippines.

Abstract

Since the first reported occurrence of *Pyrodinium* red tides in Brunei Darussalam in 1976, major red tides have recurred in 1980 and in 1988.

It is significant that during 1988 no plankton blooms were observed. However, high levels of PSP toxins (up to 5,354 µg/100 g of flesh) were recorded in the green mussel *Perna viridis* and there was a significant increase in the numbers of *Pyrodinium bahamense* var. *compressum*. Regular monitoring for the presence of *Pyrodinium* at selected stations in the coastal waters and for PSP toxin levels in *P. viridis* is carried out by the Department of Fisheries.

Introduction

More than 85% of Brunei Darussalam's small population, estimated to be around 227,000, live in the coastal areas. This, together with the high per capita fish and shellfish consumption of 40 kg/year, speaks for the country's concern on toxic red tides caused by the dinoflagellate *Pyrodinium bahamense* var. *compressum*.

Brunei Darussalam experienced its first reported red tide on 11 March 1976. This spectacular event which at first affected an area of approximately 13 km^2 is described in detail by Beales (1976). A second toxic red tide was reported in 1980 and high densities of *Pyrodinium* were found in 1988. No *Pyrodinium* cells or PSP toxins were recorded during the first five months of 1989 in Brunei Darussalam. Matdanan and Selvanathan (1984) have dealt in detail with the 1980 red tide and toxicity of fish and shellfish as well as red tide monitoring in Brunei Darussalam. Some of the implications of the 1988 red tides and a proposal for a Red Tide Action Plan were described by Matdanan et al. (1988).

Pyrodinium Red Tide Locations and Toxicity

Fig. 1 indicates the locations of the visible red tides of 1976 and 1980 while Fig. 2 indicates the sampling stations from which high densities of *Pyrodinium* were recorded in 1988. Tables 1, 2 and 3 summarize the toxicities in fish and shellfish recorded during 1976 (Beales 1976), 1980 (Matdanan and Selvanathan 1984) and 1988 (Matdanan et al. 1988), respectively.

With respect to the visibility of the red tides, it is interesting to note the comment of Matdanan and Selvanathan (1984) "Whereas the 1976 red tide manifested itself very prominently in streaks, the 1980

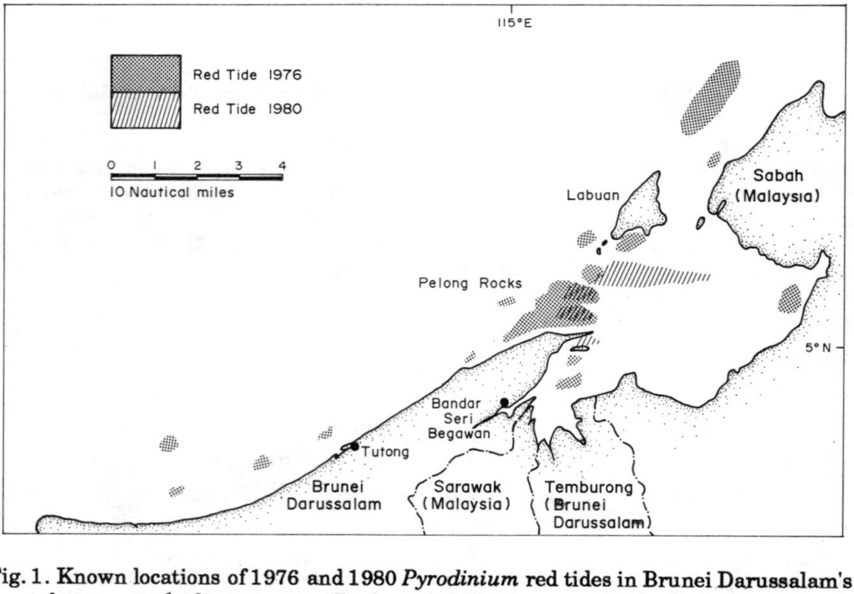

Fig. 1. Known locations of 1976 and 1980 *Pyrodinium* red tides in Brunei Darussalam's coastal waters and adjacent areas (Beales 1976; Matdanan and Selvanathan 1984).

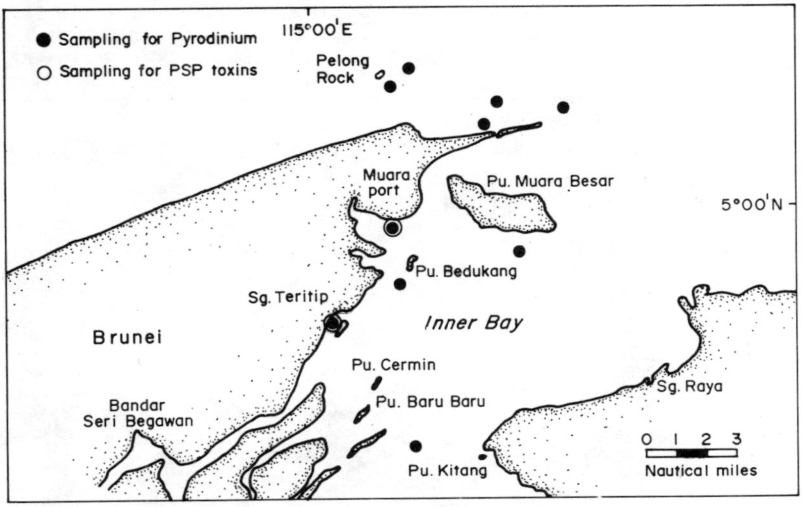

Fig. 2. Locations of sampling stations from which high densities of *Pyrodinium bahamense* var. *compressa* were recorded in vertical plankton haul samples during the 'non-visible' red tide of 1988.

occurrence was much more diffused and in most instances the organisms were not visible although they were present in samples".

The standard mouse bioassay technique (Horwitz 1970) has been used to determine PSP toxicity (but by different laboratories) during the *Pyrodinium* events of 1976, 1980 and 1988. The toxicities in *Perna viridis* of 5,354 µg/100 g of flesh on 3 February 1988 and 1,119 µg/100 g of flesh on 28 April 1988 overshadow the toxicities recorded for the visible red tides of 1976 and 1980.

This implies that:

a) Very high PSP toxicity need not necessarily be associated only with visible *Pyrodinium* red tides.
b) Regular monitoring of *Pyrodinium* and PSP toxins would be required at all times, irrespective of whether a red tide is visible or not.

Seasonality of *Pyrodinium* Red Tides

Pyrodinium red tides or toxicity in fish and shellfish due to red tides have been recorded in all months of the year (March-May 1976; April 1980-October 1981; January-September 1988), although all of them were detected during the first quarter of the year. In 1976, the

Table 1. Summary of toxic samples: red tide occurrence 1976. (Source: Beales 1976)

Date	Sample	Location	Toxicity µg/100 g meat
12 Mar	*Rastrelliger* sp. (Rumahan)	Inner Bay	99
	Sardinella sp. (Tamban)	Inner Bay	193
29 Mar	Gastropod (Tekuyong)	Sq. Bangau	71
	Gastropod (Tekuyong)	Pulau Kitang	663
	Gastropod (Tekuyong)	Tanjong Batu	876
	Gastropod (Tekuyong)	Pulau Pepatan	296
	Lamellibranch (Teritip)	Pulau Chermin	589
	Crab (Ketam)	Sg. Raya	339
	Lamellibranch (Biluyan)	Kg. Masjid Lama	847
	Lamellibranch (Tiram)	Kg. Masjid Lama	57
	Lamellibranch (Karakas)	Kg. Masjid Lama	293
27 Apr	Lamellibranch (Kunau)	Pulau Bedukang	436
	Lamellibranch (Kunau)	Pulau Muara Besar	218
	Penaeid shrimp (Udang)	Inner Bay	190
1 May	*Rastrelliger* sp. (Rumahan)	Inner Bay	478
2 May	*Rastrelliger* sp. (Rumahan)	Inner Bay	314
	Lamellibranch (Tiram)	Pulau Chermin	1,351
	Lamellibranch (Teritip)	Sg. Teritip	864
6 May	Lamellibranch (Teritip)	Sg. Teritip	2,310
12 May	Lamellibranch (Tiram)	Pulau Chermin	20
	Lamellibranch (Teritip)	Sg. Teritip	935
15 May	Lamellibranch (Tiram)	Pulau Muara Besar	274
	Lamellibranch (Biluyan)	Pulau Muara Besar	2,060

Brunei Malay names are given in brackets.

weather during the red tide occurrences was sunny and the sea calm (Beales 1976); in 1980 the days were rainy, overcast and the seas rough (Matdanan and Selvanathan 1984); and in 1988, the weather variable from sunny to rainy with variable sea conditions. A comparison of the rainfall patterns of 1987 and 1988 (Fig. 3) shows

Table 2. Summary of toxic samples; red tide occurrence 1980. (Source: Matdanan and Selvanathan 1984)

Date	Sample	Location	Toxicity µg/100 g meat
24 Jun	*Saccostrea cucullata* (Teritip)	Belangakas Jetty	52
12 Jul	*Perna viridis* (Kupang hijau)	Raft at Serasa	200
19 Jul	*P. viridis* (Kupang hijau)	Raft at Sesara	94
30 Sep	*P. viridis* (Kupang hijau)	Raft at Muara	138
6 Oct	*S. cucullata* (Teritip)	Belangkas Jetty, Muara	64
28 Oct	*Anadara granosa* (Tembayang)	Pulau Muara Besar	108
	P. viridis (Kupang hijau)	Raft at Muara	71
26 Nov	*A. granosa* (Tembayangan)	Pulau Muara Besar	83
	P. viridis (Kupang hijau)	Fisheries Station, Muara	64
4 Feb	*S. cucullata* (Teritip)	Belangkas Jetty, Muara	70
	A. granosa (Tembayangan)	Pulau Muara Besar	188
19 Feb	*S. cucullata* (Teritip)	Belangkas Jetty, Muara	66
3 Mar	*S. cucullata* (Teritip)	Belangkas Jetty, Muara	66
	A. granosa (Tembayangan)	Pulau Muara Besar	184

Table 3. Summary of toxic samples in Brunei Darussalam from January to August 1988. (Source: Matdanan et al. 1988)

Date	Sample	Location	Toxicity µg/100 g meat
23 Jan	*Perna viridis* (Kupang hijau)	GAMAFCO	77
	Selar kala (Ikan selidai)	Brunei Darussalam	51
26 Jan	*Perna viridis* (Kupang hijau)	GAMAFCO	81
	Perna viridis (Kupang hijau)	Serasa	307
	Selar mate (Temenong)	Pasar Tutong	44
	Acetes (Bubok)	Brunei Darussalam	45
	Clupeidae (Kuasi)	Brunei Darussalam	55
	Rastrelliger kanagurta	Brunei Darussalam	54

Continued

Table 3. Continued

3 Feb	*Perna viridis* (Kupang hijau)	GAMAFCO	1,054
	Perna viridis (Kupang hijau)	Serasa	5,354
2 Mar	*Perna viridis* (Kupang hijau)	GAMAFCO	64
	Perna viridis (Kupang hijau)	Serasa	86
15 Mar	*Perna viridis* (Kupang hijau)	Serasa	66
21 Apr	*Perna viridis* (Kupang hijau)	Serasa	246
	Saccostrea cucullata (Teritip)	Serasa	68
28 Apr	*Perna viridis* (Kupang hijau)	GAMAFCO	859
	Perna viridis (Kupang hijau)	Serasa	1,119
	Saccostrea cucullata (Teritip)	Serasa	487
10 May	*Perna viridis* (Kupang hijau)	GAMAFCO	430
26 May	*Perna viridis* (Kupang hijau)	GAMAFCO	69
14 Jul	*Perna viridis* (Kupang hijau)	GAMAFCO	65
	Perna viridis (Kupang hijau)	Serasa	56
3 Aug	*Perna viridis* (Kupang hijau)	GAMAFCO	46
	Perna viridis (Kupang hijau)	Serasa	45
6 Aug	*Saccostrea cucullata* (Teritip)	Anduki	53
4 Sep	*Anadara granosa* (Tambayangan)	Pulau Muara Besar	34

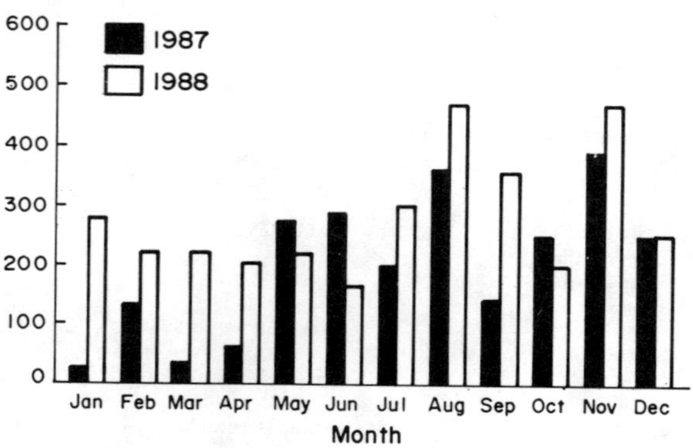

Fig. 3. Rainfall (in mm) for the years 1987 and 1988. (Source: Civil Aviation Department).

that there was more rain during the period of red tide in 1988 than during the same period in the previous year. There was no marked difference in the trend in the rainfall pattern from July to December in the two years. The rainfall pattern of the three months prior to the appearance and increasing density of *Pyrodinium* in January 1988 is comparable with that of the same period in 1989 which was followed by a *Pyrodinium*-free period. There appears to be no apparent correlation between the prevailing weather conditions and the commencement of red tides. However, one cannot eliminate the possibility of a relationship between the weather in combination with other environmental factors and the triggering of rapid reproduction or even the germination of *Pyrodinium* cysts. This could be further complicated if such triggering occurred well before the appearance of the red tide, particularly in view of the possibility of a lag period between the triggering and germination if resting cysts are implicated.

Effects of *Pyrodinium* Red Tides in Brunei Darussalam

Brunei Darussalam has been very fortunate in being able to avoid fatalities during the past *Pyrodinium* red tide occurrences. This can be attributed primarily to timely action taken by the relevant authorities to prevent the consumption of PSP toxin contaminated fish or shellfish and the cooperation of the public. At least in 1976 and in 1988, the toxin levels recorded in several shellfish exceeded manyfold the level of 80 µg/100 g of flesh recognized widely as the upper limit for safe consumption. Fourteen nonfatal cases of PSP were recorded in 1976 and most cases were prior to the identification of the red tide and subsequent action by the authorities. As pointed out by Beales (1976) the warnings given to the public and other actions might have prevented fatalities as the PSP levels recorded in some organisms would have been sufficient to cause serious illness if not death.

The red tide also left its mark in Brunei Darussalam in the form of heavy economic losses to the fishing industry. For the period 12 to 20 March 1976 alone a total of 52,578 kg of seafoods valued at $198,000 were seized and condemned (Beales 1976). In 1980, following the confirmation of public reports of red tide blooms as being due to *P. bahamense* var. *compressum* a public warning was issued on 29 April 1980 against the consumption of molluscs, particularly, mussels and other bivalves (Matdanan and Selvanathan 1984). This early action prevented any PSP-related health problems in 1980. Although economic losses were inevitable due to the public warnings, no figures are available.

On 5 December 1987, a ban was enforced on the importation and consumption of planktivorous fish such as *Sardinella* spp. and molluscan shellfish from red tide affected areas of East Malaysia even before the detection of *Pyrodinium* in the coastal waters of Brunei Darussalam. This ban and public warnings on red tides followed the discovery of high densities of *P. bahamense* var. *compressum* in the stomach contents of imported *Sardinella* spp. which had caused fatalities and sickness in a large number of cats in early December 1987 (Matdanan et al. 1988). The normal frequency of monitoring for *Pyrodinium* in the 10 pre-selected sites in the coastal waters and for PSP toxins in the green mussel *Perna viridis* by the Department of Fisheries was increased following the red tide alert. By mid-January 1988, *Pyrodinium* was observed in plankton samples and on 23 January 1988 PSP toxins were recorded in *P. viridis* (77 μg/100 g of flesh) and the scad *Selar kala* (51 μg/100 g). Although the toxin levels recorded were marginally below the recognized health limit of 80 μg/100 g of flesh, a ban on the marketing of planktivorous fish and molluscan shellfish was enforced in addition to the earlier ban on imports. On 3 February 1988, a very high PSP toxin level of 5,354 μg/100 g of flesh was recorded in wild *P. viridis* while a level of 1,054 μg/100 g of flesh was recorded in cultured *P. viridis* from the Gahasa Marine Aquaculture Farming Company (GAMAFCO) Green Mussel Farm.

The ban on harvesting of molluscan shellfish was lifted in October 1988, after one month of continuous negative results for PSP toxins in shellfish in the usual sampling and other randomly selected sites. GAMAFCO, the only operational mussel farm in Brunei Darussalam, estimated their financial losses due to the ban on harvesting as a result of the 1988 *Pyrodinium* red tide as follows:

Total overheads spent for 10 months, i.e. wages, petrol and maintenance costs, etc.		$ 31,000.00
Cost of 300 strings of imported spat (cost, freight, etc.)		$ 5,420.80
60% losses to 120,000 kg of marketable mussels due to age or dropping off string due to overgrown size		$ 75,600.00
Total loss	Brunei	$ 112,020.80

In addition to the financial losses of GAMAFCO, the importers of planktivorous fish and molluscan shellfish as well as fishermen themselves suffered financial losses during the red tide occurrence.

Public fears and consumer panic leading to the avoidance of seafood is another adverse effect of red tides. With a better understanding and upgrading of mitigation measures of red tides, the situation has improved considerably since 1976.

No fish kills have been observed during red tide occurrences in Brunei Darussalam.

At present, mitigating efforts are aimed at careful monitoring of the coastal waters for *Pyrodinium* and PSP toxins, research and education of the public. A better understanding of *Pyrodinium* red tides and the constant updating of mitigation measures since the 1976 red tide minimized PSP-related public health problems, consumer panic and losses to the fishing industry during later red tide occurrences. The ability to identify seafood liable to cause PSP has been a positive step in this direction. A detailed discussion on the monitoring of *Pyrodinium* and PSP toxins and the Red Tide Action Plan are contained in De Silva et al. (this vol.).

References

Beales, R.W. 1976. A red tide in Brunei's coastal waters. Brunei Mus. J. 3(4): 167-182.

Horwitz, W., editor. 1970. Official methods of analysis of the Association of Official Analytical Chemists. AOAC. Washington, DC.

Matdanan Haji Jaafar and S. Selvanathan. 1984. Occurrences of red tide in Brunei Darussalam and methods of monitoring and surveillance, p. 17-24. *In* A.W. White, M. Anraku and Kok-Kuang Hooi (eds.) Toxic red tides and shellfish toxicity in Southeast Asia. Southeast Asian Fisheries Development Center, Singapore and International Development Research Centre, Ottawa, Canada.

Matdanan Haji Jaafar, M.W.R.N. De Silva, P.H.Y. Sharifuddin and Siti Amin Mahali. 1988. A Red Tide Action Plan for Brunei Darussalam. Paper presented at ASEAN-US Coastal Resources Management Project Technical Workshop on Integrated Tropical Coastal Area Management, 28-31 October 1988. National University of Singapore, Singapore.

Summary of Red Tide and Paralytic Shellfish Poisonings in Sabah, Malaysia

TING THIAN MING
JOSEPH TUNG SANG WONG
Department of Fisheries
Sabah, Malaysia

TING, THIAN MING and JOSEPH TUNG SANG WONG. 1989. Summary of red tide and paralytic shellfish poisonings in Sabah, Malaysia, p. 19-26. *In* G.M. Hallegraeff and J.L. Maclean (eds.) Biology, epidemiology and management of *Pyrodinium* red tides. ICLARM Conference Proceedings 21, 286 p. Fisheries Department, Ministry of Development, Brunei Darussalam, and International Center for Living Aquatic Resources Management, Manila, Philippines.

Abstract

Red tides caused by *Pyrodinium bahamense* var. *compressum* and/or paralytic shellfish poisonings have occurred intermittently in Sabah waters since 1976. There have been over 300 reported illnesses and over 30 deaths. Most illnesses resulted from eating bivalve shellfish; the remainder from eating fish (*Sardinella* and *Decapterus* spp.). Only the intestines of the fish contained toxin.

In Kuala Penyu lagoon, the *Pyrodinium* population fluctuates with peaks in May-June and September-October. Oysters on racks in the lagoon show corresponding peaks in toxicity while cockles in the muddy bottom contain little toxin.

History of Red Tide and PSP Outbreaks in Sabah

The coastal waters along the 300 km west coast of Sabah, Malaysia, have been subject to sporadic blooms of the toxic dinoflagellate *Pyrodinium bahamense* var. *compressum* over the past twelve years. To-date numerous blooms and over 300 paralytic shellfish poisoning cases and several deaths have been reported (Table 1).

Table 1. Confirmed PSP cases in Sabah.

Date	Location	Illnesses	Fatalities	Shellfish
15-1-1976	Putatan	8	2	clams
5-3-1976	Sipitang	186	4	clams
21-3-1976	Mantanan Is.	7	1	clams
17-9-1979	Kawang	3	1	oysters
17-5-1980	Papan Is.	30	2	clams
30-12-1983	Binsuluk	9	4	*Atrina* sp.
7-1-1984	Gaya Is.	8	2	oysters
15-3-1984	Bongawan	8	5	*Oliva* spp.
6-11-1984	Sepangar Is.	5	1	*Donax* spp.
23-8-1985	Sepangar Is.	2	-	giant clam
1-10-1987	Kudat	6	-	oysters
13-10-1987	Tuaran	2	-	oysters
19-1-1988	Gaya Is.	20	3	cockles
20-1-1988	K. Kinabalu	2	-	oysters
21-1-1988	K. Kinabalu	3	-	oysters
27-5-1988	Sipitang	26	6	mussels

The first and by far the worst red tide outbreak occurred in January to May 1976 when nearly the whole 300 km stretch of the west coast of Sabah was affected. Two-hundred-and-two PSP cases and seven deaths were reported (Roy 1977).

Red tides recurred in Brunei Bay in May the following year but there was no PSP reported. In 1978 and 1979 no red tide sighting was reported, but 3 cases of PSP with one death occurred in Kawang near Kota Kinabalu in 1979 after the victims had eaten oysters collected from the beach.

There were three reported red tide sightings in 1980, one each in April, June and December, all occurring in Brunei Bay and 30 cases of PSP with 2 deaths occurred in May at Papan Island near Labuan.

Red tide resurfaced in January 1981, again in Brunei Bay. No cases of PSP were reported. In April-May 1981 red tide was reported for the first time on Sabah's southeast coast near Lahad Datu. Some fish were found dead in the vicinity where red tide had occurred. An unknown number of PSP cases was reported after the victims had eaten the shellfish *Lambis lambis,* a portion of which was tested and found very toxic. Water samples collected from the red patches contained a very dense number of *Pyrodinium* cells. The red tide slowly subsided after two weeks and has not recurred in the area since.

From 1982 to 1984 no red tide outbreak was reported. However even without visible blooms certain shellfish species continued to become toxic. During this period 30 cases of PSP with 12 deaths were recorded (See Table 1) (Wong and Ting 1984).

Red tide resurfaced on the 5th December 1985 in the vicinity of Kota Kinabalu and lasted for about a week before it disappeared after two days of heavy rainfall and bad weather. Mussels from one of the nearby islands, Sapi, were found toxic but no incidence of PSP was reported due to an advance warning issued by the Department of Fisheries.

A year later, on 9 December 1986, blooms were detected by the users of a Fisheries Department boat. Samples contained in excess of 100,000 *Pyrodinium* per liter. Several subsequent sampling trips were made in order to monitor the movement and development of the red tide patches. There were 5 to 6 patches between Gaya and Sepangar islands and the number of *Pyrodinium* cells ranged from 22,000 to 358,333 per liter. The red patches persisted for a few days. By 15 December they had disappeared though there were still a few hundred *Pyrodinium* cells per liter in the water. Meanwhile mussels from nearby Gaya and Sepangar islands had accumulated a dangerous level of toxin (in excess of 400 µg/100 g). Fortunately no PSP cases occurred due to an advance warning issued by the Fisheries and Medical Departments.

In 1987 no red tide was sighted but there were 8 cases of PSP reported, 6 in Kudat in the northern tip of Sabah and 2 in Tuaran, 30 km from Kota Kinabalu. The victims in the Kudat case ate winged oysters collected at Tigabu Island. Four of the victims received outpatient treatment while the other two victims (children) were admitted. Samples of this shellfish were found toxic (in excess of 500 µg/100 g). In the Tuaran case the victims had a meal of oysters collected from about a kilometer inside the Tuaran River mouth. A sample of the oysters was also found toxic (around 300 µg/100 g) but no *Pyrodinium* cells were detected inside the Tuaran River at the time when PSP was reported.

In January 1988, red tide broke out at two locations, Sipitang and Kota Kinabalu, 150 km apart. Several large red patches were detected in Brunei Bay and a lot of dead clams were found on the beach in Sipitang. These were analyzed and found very toxic. The public was quickly alerted and PSP averted.

Meanwhile, in Kota Kinabalu red tide was spotted on 11 January 1988. Unfortunately, despite an earlier warning, including some big headlines in the local press, the villagers living at Gaya Island picked and ate the cockles (*Anadara* spp.) on 19 January, taking advantage of the very low tide. Twenty-six people were reported poisoned, many of them seriously and three of them died. The toxin in the cockles scored over 1,400 μg/100 g. Oysters from Sembulan in Kota Kinabalu were also found extremely toxic and 5 persons were reported affected after eating them.

In March 1988, several patches of red tide were again detected near Kota Kinabulu but this time there were no PSP cases reported, partly because the public had been forewarned and no doubt partly because the memory of the recent January tragedy helped to deter them from eating shellfish.

In May 1988, large patches of red tide appeared once again in Sipitang and unfortunately had a devastating consequence in that 20 people were poisoned with 6 deaths after the victims had consumed a large meal of green mussels *(Perna viridis)* which they had collected from the "Bagang" posts (lift-net fishing stakes popular among some of the fishermen in Sabah). This happened despite earlier warning of high PSP risks in the area following the January outbreak and by the fact that significant numbers of *Pyrodinium* cells continued to be present in the area during the following months until May when again *Pyrodinium* bloomed.

Following the detection of toxin in the mussels, the said mussels have been closely monitored. They are a good indicator species and constant monitoring of their toxicity levels can give a fairly good picture of the overall red tide situation in Brunei Bay.

From June 1988 until April 1989 no red tide was detected and no PSP reported.

Paralytic Shellfish Poison in Fish

In 1987, 20 cases of PSP from ingestion of fish involving humans and several cases involving domestic animals were reported. Sardines (*Sardinella* spp.), caught in abundance around Kota Kinabalu and Papar area during the months of October and November and the round scads (*Decapterus* spp.) were implicated in the poisoning. Mouse bioassay tests carried on several sardine fish samples showed that only the fish guts were toxic and the highest toxin level recorded was 572 μg/100 g meat. Microscopic examination of the fish gut contents revealed the presence of a large number of *Pyrodinium* cells. No fish kills were reported. Following the detection of toxin in the fish, a notice was issued to the public advising them to discard the fish guts and gills and to clean the fish thoroughly before eating them. Similar incidences involving the same species of fish were also reported in Indonesia (Adnan 1984) and the Philippines (Estudillo and Gonzales 1984).

Distribution of Toxic Shellfish

The distribution of toxic shellfish conforms well with the areas affected by red tide. Results of toxin analyses in shellfish indicate that toxicity varies from place to place, from species to species and from month to month.

Toxic shellfish are found in many locations along the west coast particularly in Sipitang, Kuala Penyu, Binsuluk, Kota Kinabalu and Kota Marudu which are considered as high risk areas (See Fig. 1). For the first time since 1976 toxic shellfish were found near Tawau at the beginning of 1988, more or less at the time when red tide was reported in the neighboring Indonesian waters near Sebatik island. However, *Pyrodinium* cells were not detected in the waters around Tawau.

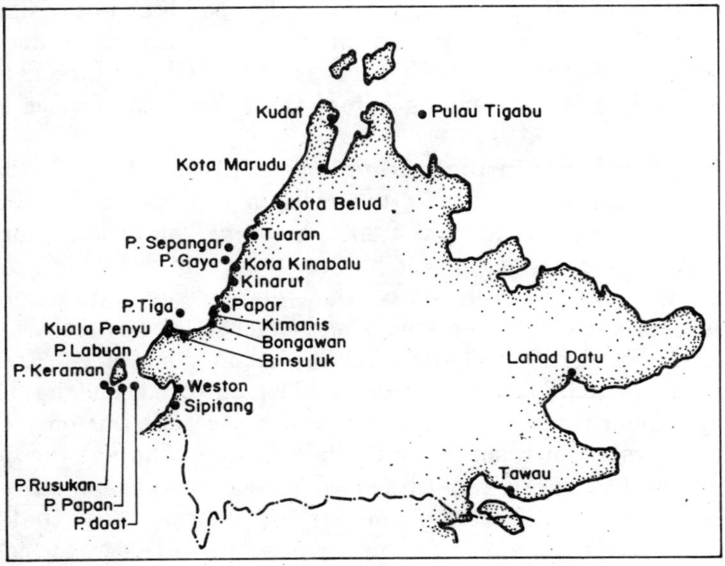

Fig. 1. Geographical distribution of toxic shellfish in Sabah.

The highest toxin level recorded to-date is 6,578 μg/100 g found in the green mussels in Brunei Bay near Sipitang. This coincided with the red tide outbreak in the Bay in May last year. Clams (*Meretrix* spp.) collected at the beach were also found toxic but the toxin level was much lower than that of the mussels.

Kuala Penyu Lagoon

Kuala Penyu lagoon represents a special case for which there is a large amount of data. This roughly S-shaped lagoon is situated near the coastal town of Kuala Penyu 120 km south of Kota Kinabalu. Oysters *(Crassostrea iredalei)* and cockles *(Anadara granosa)* are cultured in the lake by the Fisheries Department. The oysters were first found toxic in November 1983 and since then the lagoon has been closely monitored. Regular samples of the plankton are collected from the lagoon and filtered and the dinoflagellate species are identified and counted. Similarly, the toxin levels in the cultured oysters and cockles are constantly being monitored. Data collected over the past few years indicate that *Pyrodinium* blooms twice a year in the lagoon, in May to June and September to October with a peak occurring in October. Fig. 2 shows the monthly variation in the oyster toxicity and the fluctuation in the numbers of *Pyrodinium* cells for the past three years in the lagoon. The blooms are preceded by a slow rise in the cell counts; in the stable nutrient-rich waters in the lagoon the *Pyrodinium* cells can increase to 40,000 per liter but visible red patches such as those observed in the open sea have not been detected in the lagoon so far. By accurate repeated plankton samples, it is possible to issue warning before the blooms take place in the lagoon.

The toxin levels in the cultured oysters correlate well with the rise in the number of *Pyrodinium* cells in the water. The highest toxin level recorded is around 1,400 μg/100 g. The oysters normally remain toxic and dangerous for human consumption during most of the year whereas the cultured cockles only become toxic during the peak periods, usually in September and October. However the toxicity level is always significantly lower than that in the oysters. The most likely reason could be that the cockles, cultured on the muddy bottom, seldom feed on the motile *Pyrodinium* cells but only on the occasional cysts which sink onto the lake bottom whereas the oysters cultured on the racks can filter-feed directly on the motile cells. Plankton samples collected from various depths show that most *Pyrodinium* cells remain in the upper sections of the water column. When some cockles were placed on the oyster racks during the

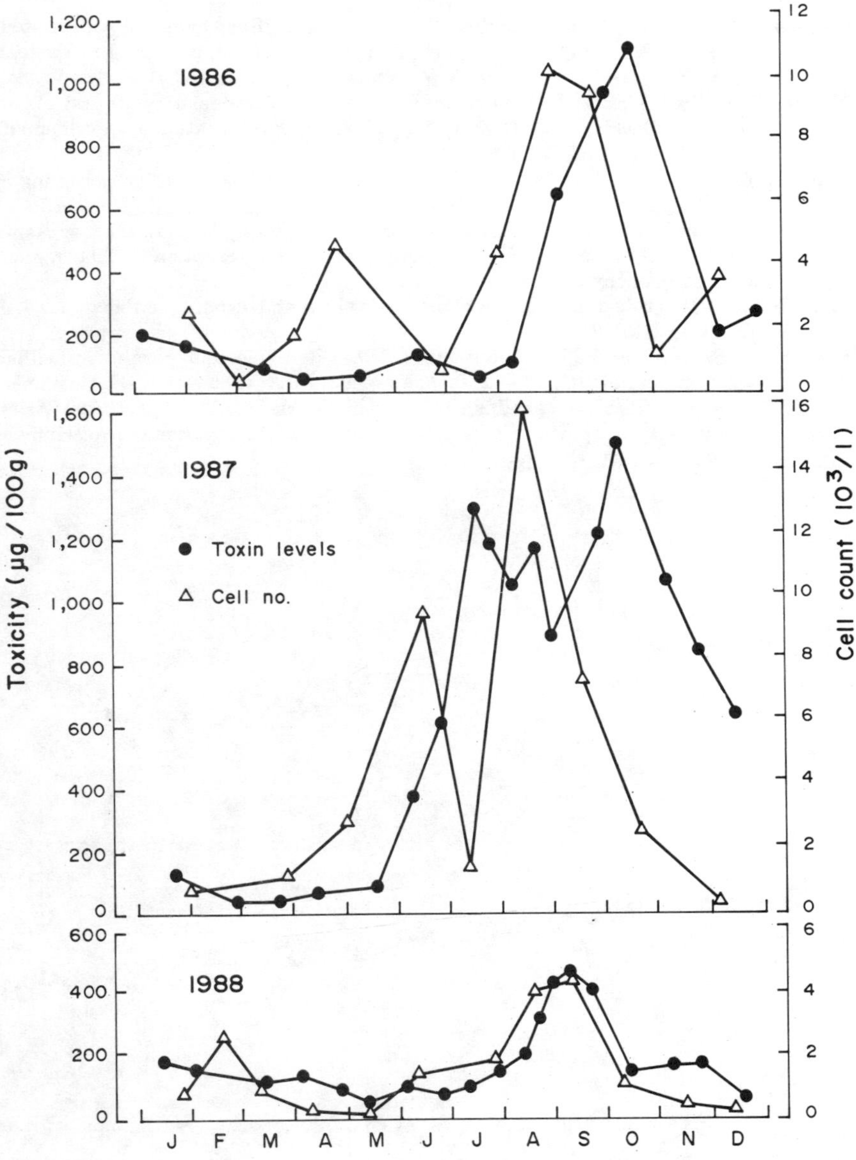

Fig. 2. Toxicities of oysters and densities of *Pyrodinium* cells in Kuala Penyu Lagoon, 1986-1988.

Pyrodinium blooms, it was found that during a short period of time the cockles had accumulated double the amount of toxin compared to the oysters.

References

Adnan, Q. 1984. Distribution of dinoflagellates at Jakarta Bay; Taman Jaya, Banten; and Benoa Bay, Bali and a report of an incident of fish poisoning in Eastern Nusa Tenggara, p. 25-27. *In* A.W. White, M. Anraku and K.-K. Hooi (eds.) Toxic red tides and shellfish toxicity in Southeast Asia. Southeast Asian Fisheries Development Center, Singapore, and International Development Research Centre, Canada.

Estudillo, R.A. and C.L. Gonzales. 1984. Red tides and paralytic shellfish poisoning in the Philippines, p. 52-79. *In* A.W. White, M. Anraku and K.-K. Hooi (eds.) Toxic red tides and shellfish toxicity in Southeast Asia. Southeast Asian Fisheries Development Center, Singapore, and International Development Research Centre, Canada.

Roy, R.N. 1977. Red tide and outbreak of paralytic shellfish poisoning in Sabah. Med. J. Malaysia 31(3): 247-251.

Wong Tung Sang, J. and Ting Thian Ming. 1984. Red tide and paralytic shellfish poisoning in Sabah, p. 35-42. *In* A.W. White, M. Anraku and K.-K. Hooi (eds.) Toxic red tides and shellfish toxicity in Southeast Asia. Southeast Asian Fisheries Development Center, Singapore, and International Development Research Centre, Canada.

Red Tides in Papua New Guinea Waters

J.L. MACLEAN
International Center for Living Aquatic Resources Management
MC P.O. Box 1501
Makati, Metro Manila
Philippines

MACLEAN, J.L. 1989. Red tides in Papua New Guinea waters, p. 27-38. *In* G.M. Hallegraeff and J.L. Maclean (eds.) Biology, epidemiology and management of *Pyrodinium* red tides. ICLARM Conference Proceedings 21, 286 p. Fisheries Department, Ministry of Development, Brunei Darussalam, and International Center for Living Aquatic Resources Management, Manila, Philippines.

Abstract

The major causative organism of red tides in Papua New Guinea is *Pyrodinium bahamense* var. *compressum*. Observations on its distribution, toxicity and biology are presented.

Introduction

Red tides were unknown in the literature in Papua New Guinea prior to 1972. During March-July 1972, seven separate instances of an unusual form of poisoning claimed the lives of three children and hospitalization of 20 adults in the Port Moresby area. Government investigators were at first refused entrance to the affected villages since villagers believed that recent spraying of DDT by Health workers against mosquitoes had infected the well which supplied the village with water.

Doctors at the Port Moresby General Hospital suspected paralytic shellfish poisoning (PSP) and thus shellfish (Rhodes et al. 1975). Investigators were finally able to collect fresh and boiled shellfish from a house in a village following poisoning fatalities there in April 1972.

The shellfish were tested for PSP by the recommended mouse bioassay (AOAC 1970) in which a mouse unit (MU) of toxin is defined as the quantity of toxin which will kill a 20-g mouse in 15 minutes. There were insufficient mice of the required strain available and no reference toxin sample. However, it was clear that the bivalves assayed were quite toxic, boiled or fresh, as was an extract of the red tide, soon discovered to be present along the coast from Port Moresby for about 100 km to the southeast (Worth et al. 1975).

Karen Steidinger of the Florida Department of Natural Resources and Max (F.J.R.) Taylor of the University of British Columbia identified the causative organism as *Pyrodinium bahamense* and later amended this to *Pyrodinium bahamense* var. *compressa* (Steidinger et al. 1980). More recently Balech (1985) noted that *compressum* is the correct Latin spelling.

The author, who was working for the Fisheries Division of the then PNG Department of Agriculture, Stock and Fisheries (DASF) on mollusc culture prospects at the time, was requested to identify the shellfish and look for red tide areas. Subsequently, I carried out a series of *ad hoc* field and laboratory investigations from June 1972 to April 1974, including two further red tide seasons.

Distribution of Red Tides and PSP

A search through DASF files, Health Department annual reports and New Guinea and Papua annual reports uncovered a number of citings of red tides and probable PSP cases. There were over 7,000 reported poisonings of all kinds each year in the New Guinea (northern half of the mainland and outer islands) region alone. Details have rarely been reported. The earliest probable PSP case on record was in 1927-28, while the first recorded red tide was in 1961. A summary of 13 separate outbreaks of red tides or PSP over 18 years to 1973 is shown in Fig. 1 (Maclean 1973). There has been no mention of further red tide research or reports in annual reports of the Fisheries Research Division since then, apart from a sighting of a red tide in Milne Bay, eastern PNG, in 1986.

The evidence suggests that red tides in PNG are irregular or in some cases, such as Port Moresby, have been annual seasonal events for a number of years. Poisonings are scattered, isolated occurrences. There were 10 known deaths and 160 treated PSP illnesses over the 18 years surveyed.

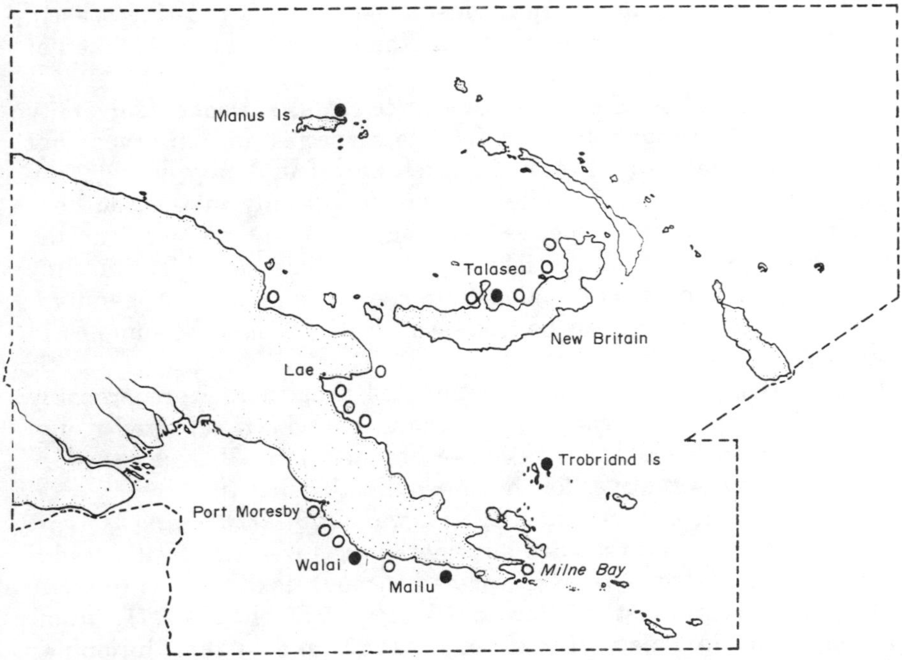

Fig. 1. Papua New Guinea showing areas in which fatal paralytic shellfish poisonings (solid circles) and red tides (open circles) have been reported. Adapted from Maclean (1973).

Red Tide Organisms

P. bahamense var. *compressum* was identified as the causative organism in surveys of red tide affected areas along the southern coast and part of the northern coast of the mainland as well as in New Britain (Fig. 1).

The only other dinoflagellate recorded as forming a red tide in PNG is *Gonyaulax polygramma* in Madang, 1969 (Maclean 1973). *Trichodesmium* blooms have been recorded from time to time.

Plankton samples were collected using a 48-micron net in various parts of PNG as time and circumstances allowed. *Pyrodinium* was found at all mainland sites from Wewak in the northwest to Yule Island, 100 km west of Port Moresby on the south coast. It was present in northwestern New Britain but not at the easterly end of that island at Rabaul. It was absent in hauls from four stations in the Trobriand Islands in May 1973 (Maclean 1977). Yet there are clear records of previous mass outbreaks of PSP in the area around this time of year (Maclean 1973). From the results of a time series of hauls in Port Moresby (see below), the absence of *Pyrodinium* in near surface plankton hauls is not unusual outside the red tide season. Not too much weight can be placed on the results of samples taken at one point of time only.

Subsamples of 68 hauls were sent to Dr. J.D. Dodge, University of London, who found 107 dinoflagellate species and observed that only 19 occurred in all areas; 6 species were found only in waters of the northern mainland, while 14 were found only in the southern mainland; a further 14 species were found only in samples from the eastern mainland and outer islands. Some differences in habitat may be assumed. The list of dinoflagellate species is given in Appendix I. Relative "abundance" can be assessed by the number of samples in which each species was present.

Dr. Dodge (pers. comm.) commented that there were probably many more species present which would have been captured using, for example, a 10-20 μm filtration technique for small *Oxytoxum* spp. and special preservatives for the naked dinoflagellates.

At the time of these investigations, *Pyrodinium bahamense* var. *compressum* was not recorded anywhere in the western Pacific ocean. However, it has since been discovered through toxic blooms to occur along the west coast of Borneo (Beales 1976; Roy 1977), in an enclosed bay in Palau (Harada et al. 1982) and in the Philippines (Hermes 1983). Further details appear in this volume. Together these occurrences reduce the value of detailed analysis of the PNG "habitat" in isolation.

Seasonality

Red tide

A list of "apparent" red tide seasons in different PNG districts was presented earlier (Maclean 1973). With regard to *Pyrodinium*-suspected blooms on the south coast, blooms have extended from November to June; on the north coast of the mainland, October to April; west New Britain, January to July; and the Trobriand Islands, April to June. These are, then, events of the northwesterly monsoon, which may begin between November and February, and wane between May and July. Nevertheless, the climate varies considerably between districts.

In the Port Moresby area, rainfall is highly seasonal and the correlation between blooms and rainfall is compelling (Fig. 2). On the north Morobe coast, rain is abundant all year, but in the highlands above it, seasonality is apparent, resulting sometimes in very marked end-of-year peaks in discharge of the large Waria River into the sea in the middle of the Morobe District (Fig. 3).

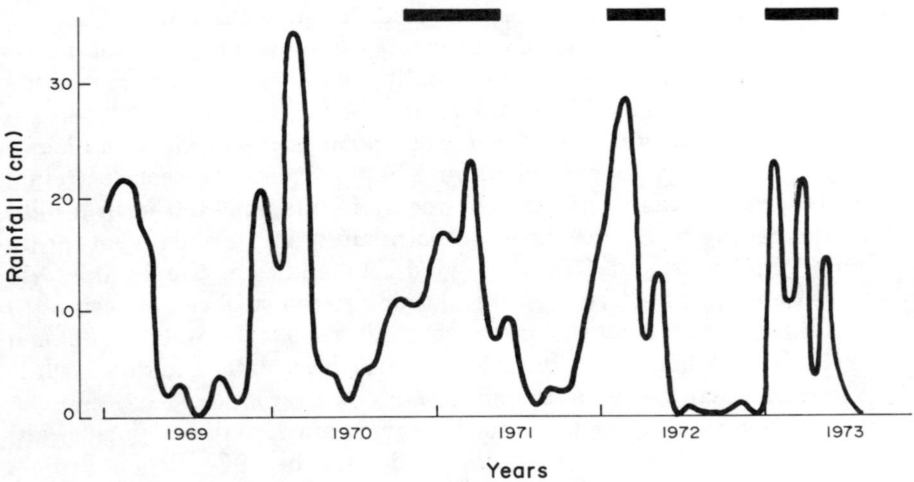

Fig. 2. Monthly rainfall (in cm) at Port Moresby, showing red tide seasons (black bars). Adapted from Maclean (1977).

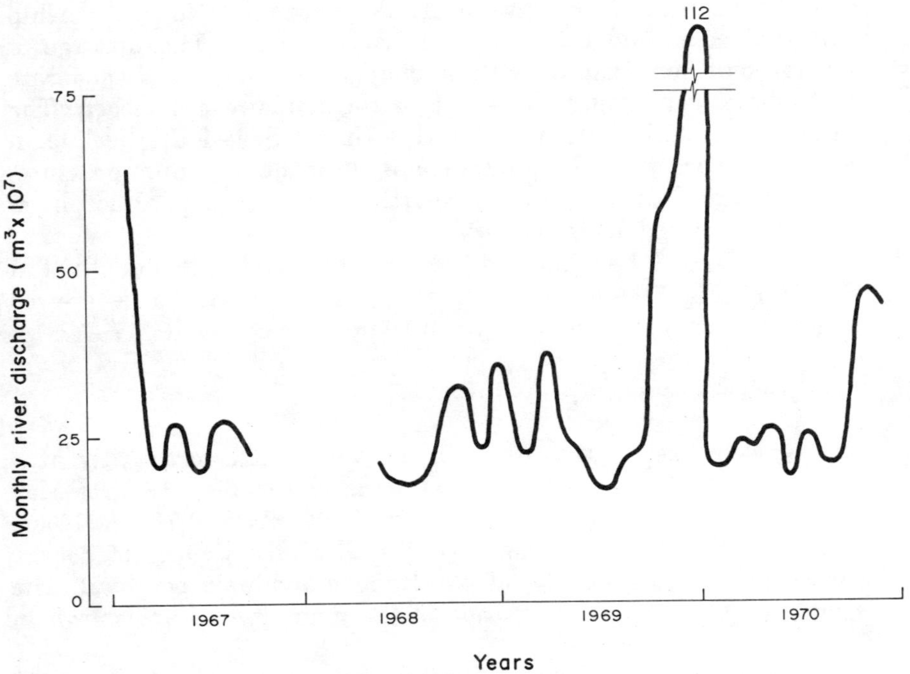

Fig. 3. Monthly river discharge (m^3 x 10^7) for Waria River, Morobe district. Adapted from Maclean (1977). There were marked increases in early 1967 and at the end of 1969.

I suspect that run-off may be important for *Pyrodinium* red tides. Yet the coastal areas involved range from low (limestone) islands to extinct volcanoes. Common features include the presence of mangroves, fringing reefs and usually some shelter from prevailing winds during blooms (Maclean 1977).

Within this broadly defined area and season, *Pyrodinium* blooms were also found to wax and wane. The 1973 red tide season around Port Moresby was monitored by weekly flights at 3-400 m in a light plane. Strong blooms occurred in both sunny and cloudy weather, in calm conditions and 15+ knot winds. In addition, the *Pyrodinium* were found to be active swimmers, rising to the surface between 0830 and 1130 hours and sinking in the afternoons between 1330 and around 1600 hours. Early rising blooms were late "sinkers", while late blooms sank early in the afternoon.

The variability of this diurnal migration remains a puzzle, but some important clues were obtained during the 1974 blooms in Port Moresby Harbor. Over a period of four weeks in March-April, the blooms gradually waned and disappeared, then suddenly reappeared stronger (and longer) than during those previous weeks. There was no rain during this phenomenon and winds were light to moderate from the southeast. Days were cloudy or sunny. No relationship between blooms and tidal patterns was apparent. The upsurge of blooms coincided exactly with a change of wind to a moderate northwestly, confirming the effect of the northwest monsoon. The pattern of declining red tides at this time followed the decline in atmospheric pressure. Red tide was absent at the minimum pressure; very strong blooms appeared shortly after as the atmospheric pressure increased again (Fig. 4).

The effect of the wind was to cause deep mixing to a depth of at least 10 m (Fig. 5). Most of the inner Port Moresby Harbor where red tides were most common is less than 10 m deep.

Species succession

Near-surface phytoplankton hauls were made fortnightly at a station in Port Moresby Harbor from the end of May 1972 to May 1973, then weekly until the end of September 1973. A 15-cm diameter, 48-micron net was used for all hauls. Counts of genera were made microscopically, of 400-1,600 individuals per haul. The results as percentage composition of the main genera are shown in Fig. 6.

The sequence in Fig. 6 begins with the end of the 1972 *Pyrodinium* blooms, which were followed by various diatoms as the dominant phytoplankters, particularly *Chaetoceros* spp. The latter remained dominant until the following *Pyrodinium* blooms, when

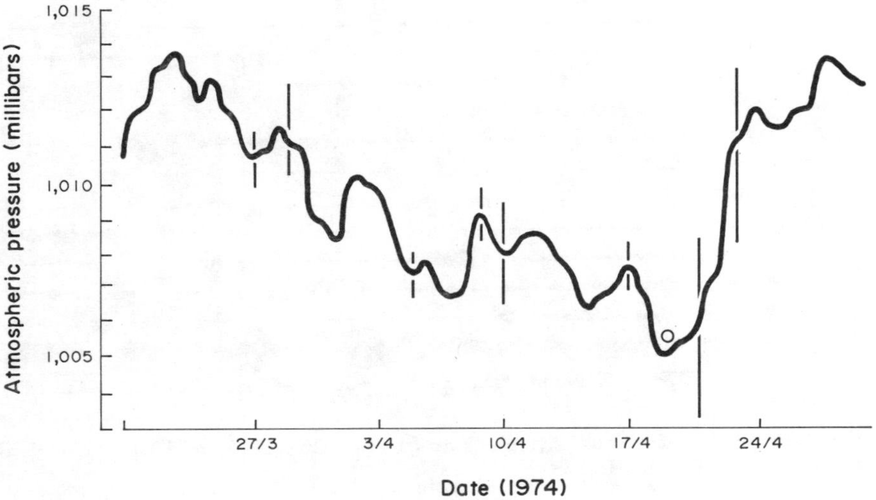

Fig. 4. Atmospheric pressure (in millibars) measured at sea level 0900 hours, March and April 1974. Relative abundance of red tide indicated by vertical bars. O - Absence of red tide. Adapted from Maclean (1977).

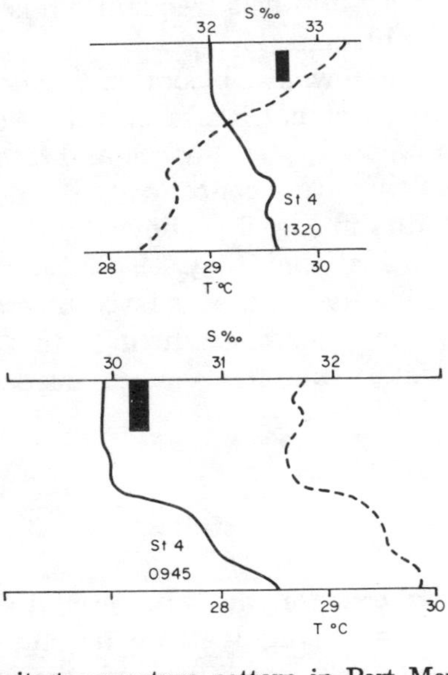

Fig. 5. Top: typical salinity-temperature pattern in Port Moresby Harbor during a southeastly wind set. A weak band of *Pyrodinium* was present at 0.5-2.0 m (29 March 1974). Bottom: influence of rainfall and wind from the northwest - reduced salinity (solid line) and reversal of the temperature gradient (dotted line) to at least 10-m deep. *Pyrodinium* formed a dense band at 0-2.5 m (23 April 1974). Adapted from Maclean (1977).

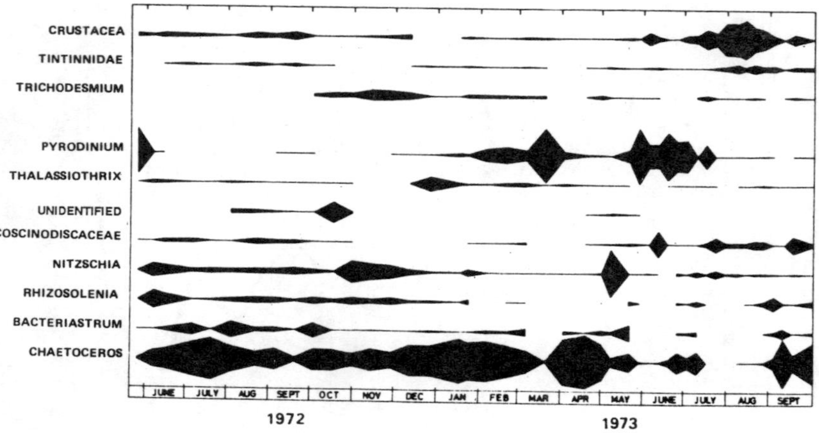

Fig. 6. Proportion of various phytoplankton genera during 1972 and 1973 at a surface station in Port Moresby Harbor.

they alternated with the dinoflagellate. A large proportion of crustaceans appeared over a two-month period following the 1973 *Pyrodinium* blooms. Thereafter, *Chaetoceros* began to dominate the hauls. Note that *Pyrodinium* was frequently absent from these hauls in the period between blooms.

Not obvious in Fig. 6 was a "bloom" of *Trichodesmium* at the end of 1972. Absent until October, rafts or colonies of *Trichodesmium* filaments appeared on the water surface and later (in January) were blown ashore, where they were concentrated in dense mats. A further limitation of the results in Fig. 6 is shown by the apparent decline in *Pyrodinium* during April and May 1973. Since hauls were made between 0900 and 1000 hours, they missed blooms which rose to the surface later in the morning during these months. Diving observations and water samples showed deeper concentrations of *Pyrodinium*.

Toxicity

Samples of *P. bahamense* var. *compressum* from plankton hauls in PNG were found to be highly toxic in mouse bioassays, while bivalve shellfish in affected areas were shown to be similarly toxic (Maclean 1975a; Worth et al. 1975).

The scattered records of PSP in the literature and the absence of recorded illness or fatality from PSP during the two red tide seasons (1973 and 1974) following the 1972 poisoning outbreak near Port

Moresby, suggest that toxicity is not a constant feature of blooms. Yet sampled shellfish in the harbor were quite toxic during the 1973 red tide there (Maclean 1975a). The Health Department had issued warnings against gathering bivalve shellfish, but gatherers were seen from time to time on the harbor foreshore in the intertidal zone.

Mortality of marine life has been associated with *Pyrodinium* blooms in PNG (Maclean 1973, 1975b), although the author did not see any mass mortality of fish during blooms.

In the Morobe blooms, the red tide seemed to have been responsible for mass mortality of oysters in an affected lagoon and reportedly for mass mortality of fish, dolphins and turtles offshore (Maclean 1975b).

The 1976 outbreak of *Pyrodinium* red tides in Sabah caused extensive kills of both pelagic and reef fish by oxygen depletion. Deep reef themselves were affected, with virtually complete faunal loss on one deep (> 6 m), sheltered reef. Shallow reefs (0-6 m) were unharmed. There was extremely low oxygen tension (< 1 ml/l) in waters deeper than 10 m during the period of highest mortality; below 10 m hydrogen sulfide was detectable and there were large numbers of decomposing organisms on deep reefs and on the bottom (Dr. E. Wood, pers. comm.). There was a scattering of several hundred dead fish in Brunei waters during this outbreak, mainly *Lethrinus* sp. (demersal) and *Stolephorus* (pelagic) (Beales 1976).

In the laboratory, I found that concentrated populations of *Pyrodinium,* resuspended from plankton hauls, killed juvenile mullet quite efficiently. The symptoms elicited were similar to those of fish in the presence of *Gymnodinium breve* (Quick and Henderson 1975). Some aspects of the data were not complete enough for formal publication but the experiments were noted in Maclean (1979). The symptoms of the mullet *in vitro* were observed in some affected fish during *Pyrodinium* blooms in the 1976 Sabah incident (Wood, pers. comm.). My experiments lacked measures of dissolved oxygen, although vigorous aeration was used to keep the cells in suspension. At this time it is not clear whether the deaths of *in vivo* pelagic fish or *in vitro* mullet were due simply to oxygen depletion or an ichthyotoxin.

References

AOAC. 1970. Official methods of analysis of the Association of Official Analytical Chemists, 11th edition. AOAC, Washington.

Balech, E. 1985. A revision of *Pyrodinium bahamense* Plate (Dinoflagellate). Rev. Palaeobot. Palynol. 45: 17-34.

Beales, R.W. 1976. A red tide in Brunei's coastal waters. Brunei Museum J. 3(4): 167-182.

Harada, T., Y. Oshima, H. Kamiya and T. Yasumoto. 1982. Confirmation of paralytic shellfish toxins in the dinoflagellate *Pyrodinium bahamense* var. *compressa* and bivalves in Palau. Bull. Japan Soc. Sci. Fish. 48(6): 821-825.

Hermes, R. 1983. Red tide: biological aspects of the Philippine case. Paper presented during the scientific session, 19 October 1983, fish conservation week. Philippine Fisheries Research Society and Bureau of Fisheries and Aquatic Resources, Quezon City, Philippines.

Maclean, J.L. 1973. Red tide and paralytic shellfish poisoning in Papua New Guinea. Papua New Guinea Agric. J. 24(4): 131-138.

Maclean, J.L. 1975a. Paralytic shellfish poison in various bivalves, Port Moresby 1973. Pac. Sci. 29(4): 349-352.

Maclean, J.L. 1975b. Red tide in the Morobe district of Papua New Guinea. Pac. Sci. 29(1): 7-13.

Maclean, J.L. 1977. Observations on *Pyrodinium bahamense* Plate, a toxic dinoflagellate, in Papua New Guinea. Limnol. Oceanogr. 22: 234-254.

Maclean, J.L. 1979. Indo-Pacific red tides, p. 173-178. *In* D.L. Taylor and H.H. Seliger (eds.) Toxic dinoflagellate blooms. Elsevier/North Holland, New York.

Quick, J.A. and G.E. Henderson. 1975. Evidence of new ichthyotoxicative phenomena in *Gymnodinium breve* red tides, p. 413-422. *In* Proc. 1st Int. Conf. Toxic Dinoflagellate Blooms, 4-6 November 1974, Boston.

Rhodes, F.A., C.B. Mills and K. Popei. 1975. Paralytic shellfish poisoning in Papua New Guinea. Papua New Guinea Med. J. 18(4): 197-202.

Roy, R.N. 1977. Red tide and outbreak of paralytic shellfish poisoning in Papua New Guinea. Papua New Guinea Med. J. 18(4): 197-202.

Steidinger, K.A., L.S. Tester and F.J.R. Taylor. 1980. A redescription of *Pyrodinium bahamense* var. *compressa* (Böhm) stat. nov. from Pacific red tides. Phycologia 19(4): 329-337.

Worth, G.K., J.L. Maclean and M.J. Price. 1975. Paralytic shellfish poisoning in Papua New Guinea. Pac. Sci. 29(1): 1-5.

Appendix I. Dinoflagellate species identified by Dr. J.D. Dodge from 68 formalin-preserved plankton hauls using a 48-micron net in various parts of Papua New Guinea, 1972-1973. Number column shows the number of samples in which the species were present.

Number of samples	
1	Blepharocysta splendor-maris
23	Ceratium breve
1	C. buceros
1	C. declinatum
1	C. deflexum
1	C. candelabrum
1	C. extensum
3	C. falcatum
54	C. furca
29	C. fusus
2	C. horridum
1	C. incisum
5	C. karstenii

Continued

Appendix I. Continued

Number of samples	
1	C. lineatum
2	C. longirostrum
1	C. lunula
2	C. macroceros
32	C. massiliense
3	C. pulchellum
1	C. teres
46	C. trichoceros
25	C. tripos
4	Ceratocorys horridum
1	Dinophysis acuta
3	D. amandula
31	D. caudata
4	D. cuneolus
2	D. doryophorum
13	D. miles
3	D. ovum
1	D. rotundata
1	D. similis
1	D. uracantha
5	Diplopeltopsis minor
21	Diplopsalis lenticula
2	Diplopsalopsis orbiculare
1	Gonyaulax diegensis
17	G. digitalis
1	G. grindleyi
6	G. monocantha
3	G. orientalis (?)
3	G. polyedra
21	G. polygramma
6	G. scrippsae
4	G. spinifera
3	G. tamarensis
1	Heteroschisma globulus
1	Noctiluca scintillans
3	Oblea rotundata
9	Ornithocercus magnificus
1	O. steinii
1	Oxytoxum scolopax
1	Pavillardinium punctata
21	Peridiniopsis (Dissodium) asymmetrica
15	Protoperidinium abei
10	P. achromaticum
1	P. avellana
2	P. breve
12	P. brochii

Continued

Appendix I. Continued

Number of samples	
15	P. cerasus
34	P. claudicans
21	P. conicum
32	P. crassipes
6	P. depressum
6	P. diabolus
20	P. divergens
9	P. excentricum
1	P. exiquipes
7	P. fatulipes
18	P. globulus
9	P. grande
5	P. granii
2	P. leonis
1	P. matzenaurii
5	P. minutum
7	P. nipponicum
2	P. nux
1	P. obtusum
2	P. oceanicum
18	P. pellucidum
12	P. pentagonum
3	P. punctulatum
1	P. pyriforme
1	P. quinquecorne
1	P. solidicorne
1	P. sphaericum
21	P. steinii
8	P. subinerme
8	P. thorianum
1	P. thulesense
1	P. tristylum
3	P. tuba
3	P. wiesneri
4	Podolampis bipes
1	Prorocentrum compressum
2	P. gracile
7	P. maximum
3	P. micans
4	P. sigmoides
1	Pyrocystis lunula
2	P. pseudonoctiluca
1	P. robusta
47	Pyrodinium bahamense
16	Pyrophacus horologium
6	P. steinii
4	Scrippsiella trochoidea
5	Triadinium polyedricum

Pyrodinium Blooms and Paralytic Shellfish Poisoning in the Philippines

CIELITO L. GONZALES
Bureau of Fisheries and Aquatic Resources
860 Quezon Avenue, Quezon City
Philippines

GONZALES, C.E. 1989. *Pyrodinium* blooms and paralytic shellfish poisoning in the Philippines, p. 39-47. *In* G.M. Hallegraeff and J.L. Maclean (eds.) Biology, epidemiology and management of *Pyrodinium* red tides. ICLARM Conference Proceedings 21, 286 p. Fisheries Department, Ministry of Development, Brunei Darussalam, and International Center for Living Aquatic Resources Management, Manila, Philippines.

Abstract

Pyrodinium blooms and associated paralytic shellfish poisoning (PSP) have occurred in the Philippines in 1983, 1987 and 1988-1989. The first event was in the central Philippines in Samar where there were 278 reported cases of PSP. The phenomena were repeated in 1987. This time, red tides of *Pyrodinium* and PSP were also present in the northwestern Philippines in Zambales. In 1988-1989 an outbreak occurred in Manila Bay, as well as in Samar, where it was more widespread than on the previous occasions. Negros Occidental, Capiz and Cebu Provinces were affected for the first time during this outbreak.

Introduction

Red tide was reported in the Philippines as early as 1908 (Smith 1908). The people of Bataan Province suspected that associated fish kills were the effect of the dumpings from the sanitary barge "Pluto". An investigation conducted by the United States Fish Commission steamer "Albatross" proved that the mortality among fishes was due to *Peridinium* blooms in Manila Bay. In recent years, minor nontoxic red tide outbreaks became almost an annual event in Manila Bay, particularly in the Cavite area. It was not until June 1983 that the first outbreak of a toxic red tide occurred, in Samar, Central Philippines.

This paper discusses the events related to the *Pyrodinium* blooms in the Philippines, up to mid-1989. Toxic blooms occurred in 1983, 1987, 1988 and 1989. Locations mentioned are shown in Figs. 1-3.

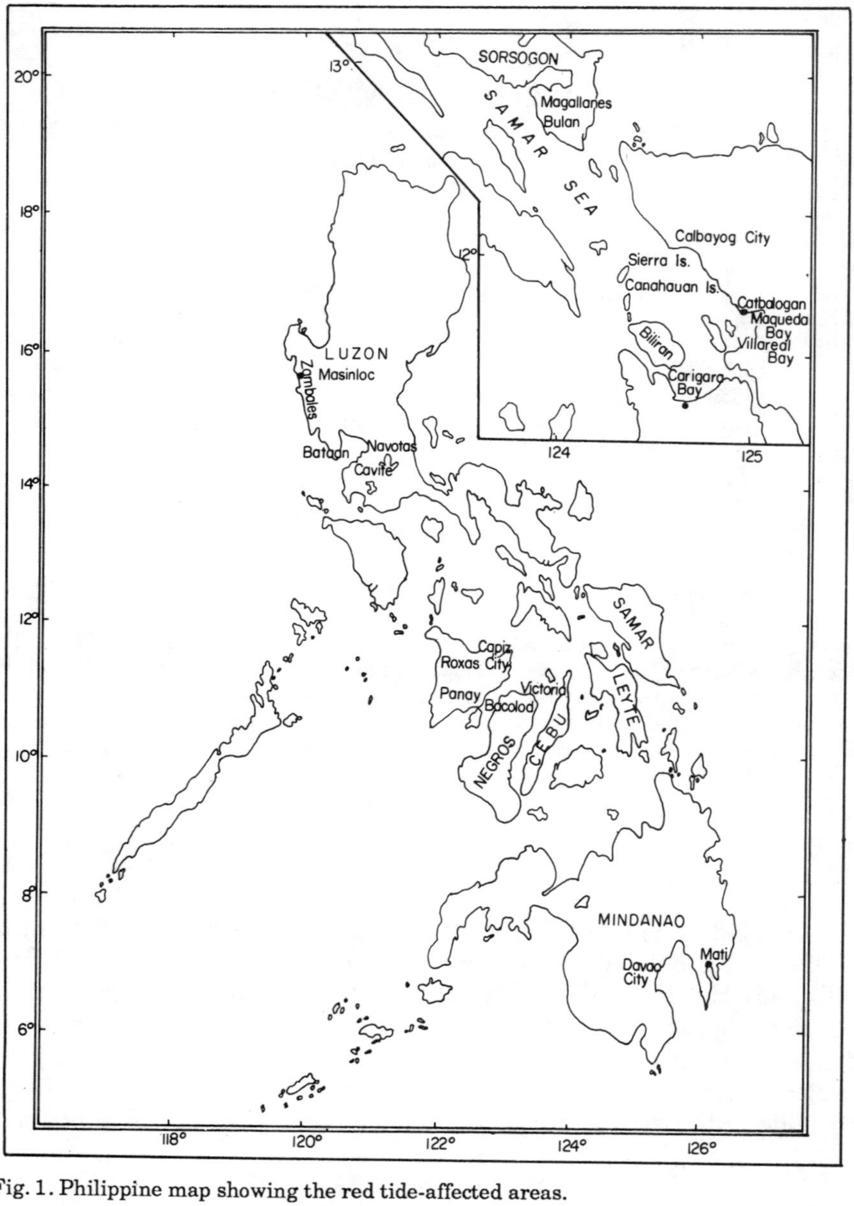

Fig. 1. Philippine map showing the red tide-affected areas.

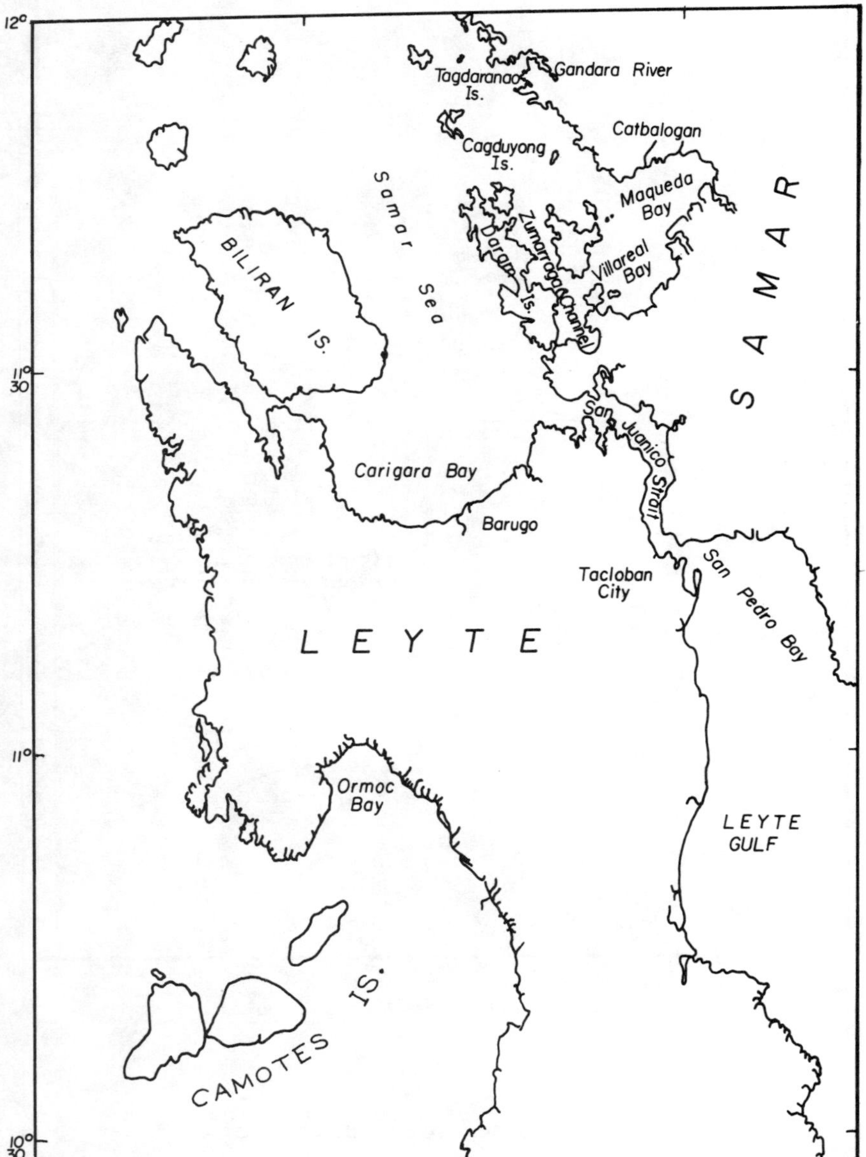

Fig. 2. Map of Samar and Leyte.

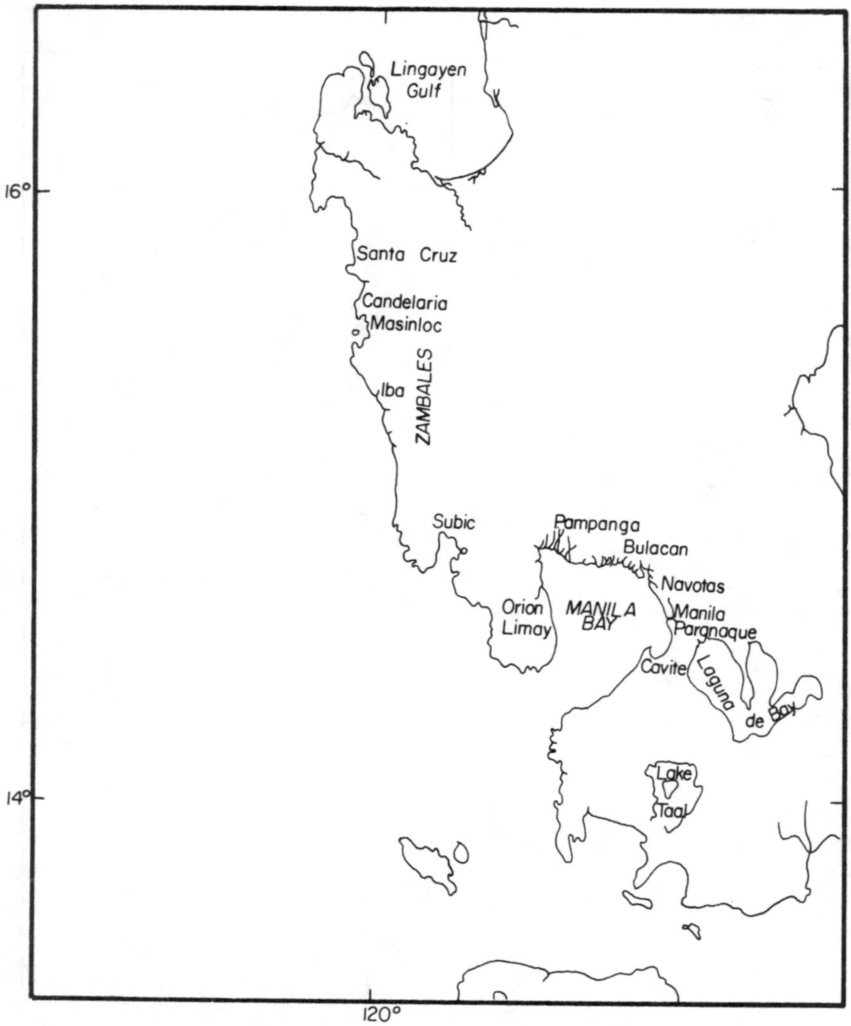

Fig. 3. Map of Luzon showing the Zambales and Manila Bay areas.

The 1983 Red Tide Occurrence

The first recorded paralytic shellfish poisoning (PSP) cases in the Philippines occurred in Catbalogan, Samar, on 21 June, involving a family of eight. Two children, aged 3 and 7 years, died ten hours after eating boiled green mussel *(Perna viridis)*. Other members of the family became ill and were sent to the hospital. One young boy who did not consume the shellfish was not affected. The symptoms were those of PSP. Twelve other families in the neighborhood who also had green mussels for dinner suffered the same symptoms (Estudillo 1983; Nabong-Cabardo 1983).

Specimens of stomach and intestinal contents sent to the National Bureau of Investigation and the Ministry of Health in Tacloban City failed to show any bacterial pathogens or chemical poison in the specimens.

The number of poisoning cases continued to increase. In Catbalogan alone, a resident physician at the provincial hospital reported that as many as 30 poisoning victims were being admitted in a single day. In the wake of illnesses and deaths caused by the consumption of green mussels, a team of researchers was sent by the Bureau of Fisheries and Aquatic Resources (BFAR) to the area on 8 July. The plankton samples confirmed a dinoflagellate bloom in Maqueda and Villareal Bays.

The organism was identified by Prof. Rudolf Hermes of the Philippine-German Fishery Project, University of the Philippines, as *Pyrodinium bahamense* var. *compressa* (Böhm 1931). The identification was confirmed by J.L. Maclean of the International Center for Living Aquatic Resources Management (ICLARM) who made an extensive study on *Pyrodinium* blooms in Papua New Guinea, and by B. Roberts of the Marine Research Laboratory, Department of Natural Resources, Florida (Estudillo 1983). Later, Matsuoka et al. (1989) proposed that the name of variety should be spelled as *compressum* in accordance with the International Code of Botanical Nomenclature.

The visible red tide in Maqueda Bay spread out to other areas up to Calbayog City, apparently due to a tropical storm which occurred from 13 to 15 July.

A month later, the blooms were no longer visible in Samar Sea, except in the areas southwest of Tagdaranao Island, north of Calbayog City, and near the mouth of Gandara River. The water discoloration completely disappeared in Maqueda Bay, Villareal Bay and Samar Sea after a week (Estudillo and Gonzales 1984).

In late July, 49 cases of PSP with one death, attributed to the ingestion of green mussels, were reported from Bulan, Sorsogon, approximately 185 km northwest of Maqueda Bay. A patch of discolored water along the coast of Magallanes, Sorsogon, was observed in the third week of August.

On August 26, a single case of paralytic shellfish poisoning which was traced to the ingestion of green mussel collected from Balete Bay in Mati, Davao Oriental, some 280 km south of Maqueda Bay, was reported in Davao City (Gacutan et al. 1985). The water of

Balete Bay showed the presence of *Pyrodinium* cells at a relatively low cell density of 16,000/l).

On September 23, eight cases of PSP with one death were reported from Barangay Barra, Roxas City. The poisoning was attributed to the consumption of Asian moon scallops *(Amusium pleuronectes)* cooked in vinegar by the victims. Examination of the stomach contents of the shellfish collected from Panay waters showed the presence of *Pyrodinium* but only in small quantity (1 to 20 cells per scallop).

By the end of August, the toxic dinoflagellate had totally disappeared from Maqueda Bay, Villareal Bay, Carigara Bay and the Samar Sea, but was succeeded by blooms of *Noctiluca scintillans.* The incident left a total of 278 reported PSP cases with 21 deaths. This figure may be an underestimate considering that people living in remote villages and neighboring islands did not bother to report poisoning cases in their areas.

The 1987 Red Tide Occurrences

In 1987, an officer from the BFAR mussel farm in Masinloc, Zambales, who was previously trained in the methods of detection and identification of toxic red tides, reported to the BFAR Fisheries Research Division in Quezon City two cases of PSP involving the farm caretaker and his wife. A survey team dispatched to the area confirmed the presence of dinoflagellate blooms caused by *Pyrodinium bahamense* var. *compressum* in the coastal waters of Zambales, extending from Subic to Santa Cruz. Almost simultaneously, the toxic red tide recurred in Samar.

In Zambales, the organism was present in the waters from Subic to Santa Cruz but the discolored waters were observed only in the areas of Masinloc, Santa Cruz and Subic. It became clear that in the past two years, poisoning cases resembling PSP were admitted in the provincial and district hospitals in Zambales but due to the lack of knowledge on red tide and PSP by local physicians, the illnesses were diagnosed as simple cases of food poisoning. The total number of PSP cases in Zambales during the 1987 red tide occurrence was not known; however, the Integrated Provincial Health Office in Zambales reported that nine suspected red tide victims were admitted in the hospital from April 7 to July 27, while the Candelaria District Hospital reported four cases. It was also learned that a three-year old boy from Bamban, Masinloc, died several hours after eating a blood cockle (*halaan; Anadara* sp.). The toxic dinoflagellate started to dissipate in the coastal waters of Zambales in August (Gonzales et al. 1989).

In Samar, the greater concentrations of *Pyrodinium* were observed in the areas between Biliran and Canahauan Islands and in Carigara Bay where cell densities of more than 2 x 10^6/l and 8.57 x 10^6/l, respectively, were recorded. The densities of the toxic dinoflagellate inside Maqueda and Villareal Bay were relatively low except in the northern part of Maqueda Bay where a maximum cell density of 43,500/l was recorded (Gonzales et al. 1989). The 1987 *Pyrodinium* bloom in Western Samar lasted for three months and similar to the 1983 occurrence, the bloom was followed by an increase in the population of *Noctiluca scintillans.*

A total of 211 cases of PSP, six of which were fatal, were recorded in Samar from 26 May to 7 August and most of the victims were children of 5 to 9 years. About 67 per cent of the total PSP cases in Samar were caused by consumption of green mussel.

The 1988-1989 Red Tide Occurrences

In August 19, 1988, nine PSP cases were reported from Orion, Bataan, followed by another 28 cases in Limay within a four-day period. Plankton samples from the area revealed that there was an unusual number of *Pyrodinium* cells in the waters of Bataan and later in other areas of Manila Bay. The toxicity analysis of green mussel samples from Limay showed a very high toxin level of 1,005 µg toxin per 100 g of shellfish meat.

More cases of PSP were reported in the succeeding days resulting in a red tide scare in Metro Manila and the neighboring areas of Cavite, Bulacan, Pampanga and Bataan. A total of 121 PSP cases were reported from Bataan, Navotas and Cavite City during the period 19 August to 30 September, of which only 65 cases were validated by the Department of Health as PSP cases because the epidemiological investigation was not completed and strict, narrow criteria were being followed by the health agency in the validation of cases.

Two months after the first PSP cases were reported in Manila Bay, the toxin level in green mussel samples collected from Parañaque, Navotas and Cavite became zero. Samples from Bataan remained toxic until the Department of Agriculture decided in December to harvest the remaining contaminated mussels from Limay and Orion, Bataan. The temporary ban on shellfish which was imposed at the start of the red tide occurrence in Manila Bay was officially lifted on the third week of December.

Blooms of toxic *Pyrodinium* were experienced for the third time in Western Samar in September 1988. The bloom was spotted as a one-hectare reddish discoloration southwest of Cagduyong Island on

22 September where it remained stationary until the end of September. Then it dispersed northwestward to Calbayog City waters and southward to Zummaraga and Daram Islands. The water discoloration was reported to have affected Carigara Bay, particularly the area of Barugo, Leyte, in early November. The reddish water remained in Barugo until the end of November, after which, it moved along the inner part of the bay towards Biliran Strait where the organism was present until March 1989, although in low quantities and the water discoloration was no longer visible.

The toxic dinoflagellate spread out to San Pedro Bay in January, possibly due to the movement of water masses from Carigara Bay to Leyte Gulf, passing through the San Juanico Strait.

After a month, a big area in Ormoc Bay was also affected by the *Pyrodinium* red tide. The red tide area was estimated by the Red Tide Monitoring Team of the Department of Agriculture Regional Office No. 8 to be approximately 300 hectares.

Forty-five PSP cases including six deaths were recorded from Carigara Bay area by the Department of Health Regional Office in Tacloban City, seven cases with one death in Leyte Gulf and San Pedro Bay, and 25 cases with one death in Western Samar.

In the early period of the northeast monsoon season in 1988, an outbreak of PSP was recorded in Negros Occidental and Capiz provinces, Central Philippines. There were 277 reported poisoning cases but only 109 cases were validated as PSP. In most cases, green mussel was the transvector.

Four days after the first cases occurred, the BFAR collected plankton samples from 15 stations in the northern coastal waters of Negros Occidental which were found positive for *Pyrodinium* with cell densities ranging from 1 to 35 per liter.

On February 1, 1989, eight persons became ill and two died in Cebu province after they consumed shellfish (locally called "tagnipis") gathered from a drifting bamboo pole in Barangay Caubian, Lapu-Lapu City. A week later, two cases, exhibiting signs and symptoms of PSP, were admitted at the Lapu-Lapu District Hospital. At this point, epidemiological investigation was carried out by the Department of Health which recorded a total of 24 cases with five deaths.

Samples of scallops were collected from the area and submitted to the Bureau of Food and Drugs for mouse bioassay tests. The results of analysis showed a very high toxin content of 2,000 µg per 100 g of shellfish meat. The epidemic lasted for two weeks (Pastor et al. 1989).

Economic Implications

As a result of the red tide scare, the prices of fish and invertebrates dropped from the usual price of about 800 pesos (US$38) per tub of 35 kg to about 200-300 pesos (US$9.52-14.28) per tub (Felix 1988). Even prices of those fishes caught from other areas which were free from red tide contamination were affected by the red tide scare.

Commercial fishing boat operators had an estimated loss of about 17 million pesos (US$809,524) in a four-day period during the early part of the red tide season in Manila Bay. It was estimated that the loss incurred by the shellfish industry alone would amount to a minimum of 50 million pesos or approximately US$2.38 million (Robles 1988).

References

Estudillo, R.A. 1983. Red tide. BFAR Fish. Newsl. 12(3-4), p. 2-12.

Estudillo, R.A. and C.L. Gonzales. 1984. Red tides and paralytic shellfish poisoning in the Philippines, p. 52-79. *In* A.W. White, M. Anraku and K.K. Hooi (eds.) Toxic Red Tides and Shellfish Toxicity in Southeast Asia, Southeast Asian Fisheries Development Center/International Development Research Centre.

Felix, R. 1988. Big fishers hurt by red tide, too. Manila Chronicle. September 22, 1988:15.

Gonzales, C.L., J.A. Ordonez and A.M. Maala. 1989. Red tide: the Philippine experience, p. 45-48. *In* T. Okaichi, D.M. Anderson and T. Nemoto (eds.) Red Tides: Biology, Environmental Science and Toxicology. Elsevier, New York.

Matsuoka, K., Y. Fukuyo and C.L. Gonzales. 1989. A new discovery of cysts of *Pyrodinium bahamense* var *compressum*, p. 301-304. *In* T. Okaichi, D.M. Anderson and T. Nemoto (eds.) Red Tides: Biology, Environmental Science and Toxicology. Elsevier, New York.

Nabong-Cabardo, C. 1983. When the sea turned red. Philipp. Panorama 12(37): 12, p. 14-17.

Pastor, N.I.S., I.L. Gopez, M.C. Quizon, N.B. Bautista, Jr., F.M. White, V.S. Lofranco, R.M.S. Benabaye, M.E. White and M.M. Dayrit. 1989. Epidemics of paralytic shellfish poisoning in the Philippines, 1988-1989, lessons in science and public policy. 14 p. mimeo.

Robles, A.C. 1988. Red tide: ignorance takes its toll. Manila Chronicle, September 25, 1988: 9.

Smith, A.M. 1908. *Peridinium* (Editorial). Philipp. J. Sci. Ser. A3(3): 187-188.

The Guatemalan Experience with Red Tides and Paralytic Shellfish Poisoning

F. ROSALES-LOESSENER
DITEPESCA
Guatemala

ROSALES-LOESSENER, F. 1989. The Guatemala experience with red tides and paralytic shellfish poisoning, p. 49-51. *In* G.M. Hallegraeff and J.L. Maclean (eds.) Biology, epidemiology and management of *Pyrodinium* red tides. ICLARM Conference Proceedings 21, 286 p. Fisheries Department, Ministry of Development, Brunei Darussalam, and International Center for Living Aquatic Resources Management, Manila, Philippines.

Abstract

The first outbreak of paralytic shellfish poisoning in Guatemala occurred on the Pacific coast in July 1987 accompanied by large blooms of *Pyrodinium bahamense* var. *compressum*. There were 187 cases, of which 26 were fatal. It was recommended to carry out a permanent monitoring program of shellfish toxicity levels and initiate a public education program.

Introduction

Guatemala is a small tropical country, located in Central America. It borders Mexico to the north, Belize to the northeast and El Salvador and Honduras to the southeast. It has coasts in both the Pacific Ocean and the Caribbean Sea in the Atlantic.

In Guatemala, red tides in general and PSP in particular were completely unknown phenomena until 4 and 2 years ago, respectively. These phenomena may have occurred previously in the country, but due to lack of knowledge in this field and the absence of human poisonings, they were not reported. The first documented red tide in Guatemala occurred in August 1985, at Puerto Quetzal on the

Pacific coast, which caused water discoloration and fish kills. The micro-organism responsible was not identified. No human intoxications were involved. The second and last red tide occurred in July 1987, also in the Pacific coast of the country, causing a severe and tragic outbreak of 187 PSP cases of which 26 were fatal (Rosales-Loessener et al. 1989). The toxigenic micro-organism responsible was identified as the dinoflagellate *Pyrodinium bahamense* var. *compressum* and the toxic clam ingested by humans was identified as *Amphichaena kindermanni.* Victims were members of local fishing communities that had consumed clam chowder.

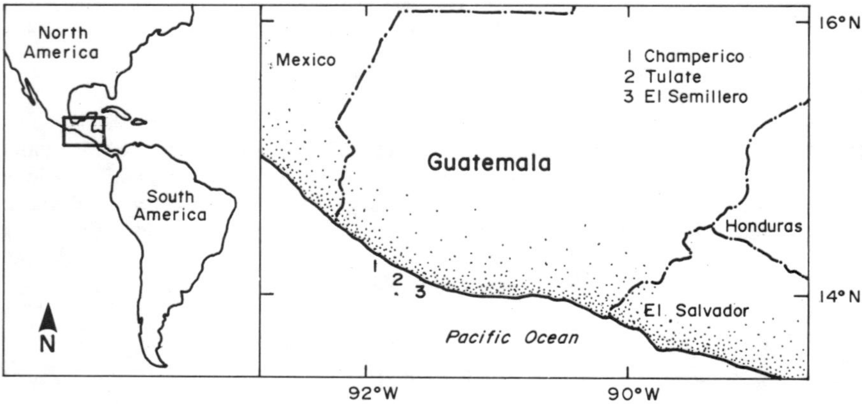

Fig. 1. Guatemalan localities implicated in paralytic shellfish poisoning, August 1987.

The 1987 Red Tide

The 187 cases of PSP were detected in five localities along a 60-km stretch of coast. In Champerico there were 21 deaths and in Finca La Verde 5 deaths during the same day. Some 50% of the mortality occurred in children aged 0-6 years, with only 5% in persons above 18 years. The main symptoms reported by several local residents after eating clam chowder were: numbness of the lips, face, fingers and toes; weakness; headache; and respiratory disorders.

In Champerico, severe symptoms appeared on 30 July. The epidemic was first recognized at about 1400 hours when patients began arriving at the local clinic. Numbers peaked at 1700 hours and the intoxication appeared to be related to ingestion of clam chowder for lunch.

Aerial surveys, carried out until 4 August, located a reddish band about 4 km wide and 40 km long parallel to the coast east from Champerico. Fishing boats in this area reported unusually warm water (about 30°C) in the red tide zone.

HPLC analysis by the US Food and Drug Administration of phytoplankton extracts, clam extracts and stomach contents from the victims, established that the primary toxin involved was 21-sulfosaxitoxin (fraction B-1), accompanied by relatively small amounts of N-1-hydroxysaxitoxin (neosaxitoxin) and saxitoxin.

As soon as it was clear that seafood was implicated in the poisoning, a moratorium was placed on all seafood harvesting. When the true nature and scope of the toxicity was recognized, the closure was limited to bivalves. Since a prolonged general closure could have caused great economic and social hardship, a great deal of emphasis was placed on promptly identifying and communicating to the public the true scope of the problem. However, there has been no detailed evaluation of the actual economic impact of these measures.

After this tragic experience, two main recommendations were given:

- To carry out a permanent monitoring program on PSP levels of bivalves from the Pacific coast of Guatemala.
- To develop a public education program on PSP in the coastal villages, by medical personnel, to alert local residents to the danger of this phenomenon.

Management Issues

When the massive intoxication was reported, the Public Health Ministry and the Agriculture Ministry established an inter-institutional commission to take some measures to control the situation, which the local press attributed to intoxication by pesticides. Also the Ministry of Public Health, through the Panamerican Health Organization, acquired the professional services of Dr. Sherwood Hall to evaluate the outbreak and to provide recommendations.

In the same way Dr. D. Rodriguez and Dr. R. Etzel from the Centers of Disease Control, Atlanta, Georgia, USA, arrived in Guatemala two weeks after the outbreak to evaluate epidemiological aspects. These experts provided to the commission basic knowledge on red tides and PSP and encouraged the commission members to continue studying this phenomenon.

Reference

Rosales-Loessener, F., E. De Porras and M.W. Dix. 1989. Toxic shellfish poisoning in Guatemala, p. 113-116. *In* T. Okaichi, D.M. Anderson and T. Nemoto (eds.) Red Tides: Biology, Environmental Science and Toxicology. Elsevier Science Publishing, New York.

Mechanisms for Red Tides of *Pyrodinium bahamense* var. *compressum* in Papua New Guinea, Sabah and Brunei Darussalam

H.H. SELIGER
Department of Biology
The Johns Hopkins University
Baltimore, Maryland 21218

SELIGER, H.H. 1989. Mechanisms for red tides of *Pyrodinium bahamense* var. *compressum* in Papua New Guinea, Sabah and Brunei Darussalam, p. 53-71. *In* G.M. Hallegraeff and J.L. Maclean (eds.) Biology, epidemiology and management of *Pyrodinium* red tides. ICLARM Conference Proceedings 21, 286 p. Fisheries Department, Ministry of Development, Brunei Darussalam, and International Center for Living Aquatic Resources Management, Manila, Philippines.

Abstract

Mechanisms are proposed to explain the development of red tides of the toxic dinoflagellate *Pyrodinium bahamense* var. *compressum* along the northern and southern coasts of Papua New Guinea and along the northwest coast of Borneo, from Sabah to Brunei Darussalam. The "spread" of red tides to areas not previously affected and the increased frequency of annual occurrences of paralytic shellfish poisoning in already colonized areas are suggested to be the result of coastal hypertrophication. Increased nutrients along coastlines permit sufficient reproduction of dinoflagellates in entrained surface patches to compensate dilution losses, enhancing emigration over longer distances.

Introduction

Shellfish toxicity and red tides of dinoflagellates are natural phenomena in many parts of the world and were present before major human settlement and modification of the environment. However during the past two decades there has been an increased frequency of annual incidences of Paralytic Shellfish Poisoning, PSP, and dinoflagellate red tides have spread to new geographical areas.

Mechanisms for the transport and retention of *Pyrodinium* blooms described for the Caribbean (Seliger et al. 1970) are applied to explain the first occurrences of red tides along the northern and southern coasts of Papua New Guinea and of northwest Borneo. It is proposed that increased discharges of industrial and human wastes into major waterways, coupled with decreased natural filtering in the watersheds due to deforestation and filling-in of wetlands have markedly increased the nutrient concentrations and the sediment loads delivered in runoff to estuarine and coastal waters. These nutrients now permit sufficient reproduction of dinoflagellates to compensate organism dilution losses, enhancing the probabilities of successful alongshore emigration.

Occurrences and Mechanisms

Surface patches of red tide dinoflagellates become visible in bright sunlight when cell concentrations approach 10^6/l. If these dinoflagellates are bioluminescent their presence at night will produce visible luminous wakes behind ships at concentrations of 10^3/l and the bioluminescent color can be identified as blue at concentrations of 10^{-4}/l. PSP due to shellfish can occur when dinoflagellate concentrations are not sufficient to produce red tides, ≈ 10^4/l (Dale et al. 1978; Anderson and Morel 1979; Lewis et al. 1979; Yentsch and Mague 1979). If their transport pathways or if the areas to which they may be delivered are identified, a minimum number of automated buoy stations can be set up to measure mechanically stimulable bioluminescence (Seliger and McElroy 1968; Seliger et al. 1969, 1970) and give advance warning of blooms by several weeks.

In an embayment or estuary, dinoflagellates can signal their presence by phototactic accumulations along convergence zones of estuarine or tidal shear or plume fronts (transient increases in surface concentrations by factors of 10-100) to form visible surface patches (10^6-10^7/l). One of the difficulties in correlating experimental observations of dinoflagellate red tides to the mechanisms by which wind and water circulation patterns produce them is that species other than dinoflagellates may also produce "red tides". Anecdotal reports of "red tides" without species identification may therefore be confusing to anyone attempting to correlate climatic and hydrographic events to the development of specific dinoflagellate red tides. For example, Maclean (1973, this volume) reports that in 1972, the year of major red tides of *Pyrodinium bahamense* var. *compressum* along the southern coast of Papua New Guinea, similarly colored streaks of the sea salp *Thalia* sp. were also identified in Port Moresby harbor and inside the barrier reefs southeast of Port

Moresby. Early in 1973 the filamentous blue-green alga *Trichodesmium* sp. also produced orange-red bands in Port Moresby harbor. Discolored water streaks or patches can also be produced when the diatom *Melosira* is stirred up from bottom sediments (Maclean 1973) or when sediments are delivered into embayments by tributary rivers (Beales 1976).

Papua New Guinea

A chart of the Papua New Guinea (PNG) region (5-10°S 145-155°E) is shown in Fig. 1 in which the areas of red tide observations during 1971-1974 are shaded. Port Moresby harbor is surrounded by mangroves and has no rivers emptying into it. On the northern side of PNG the Warier River empties into Morobe Lagoon. The monsoon season and NW winds occur from January through May; the dry season and SE winds occur from May through December. The data are abstracted from published reports and conversations with J.L. Maclean.

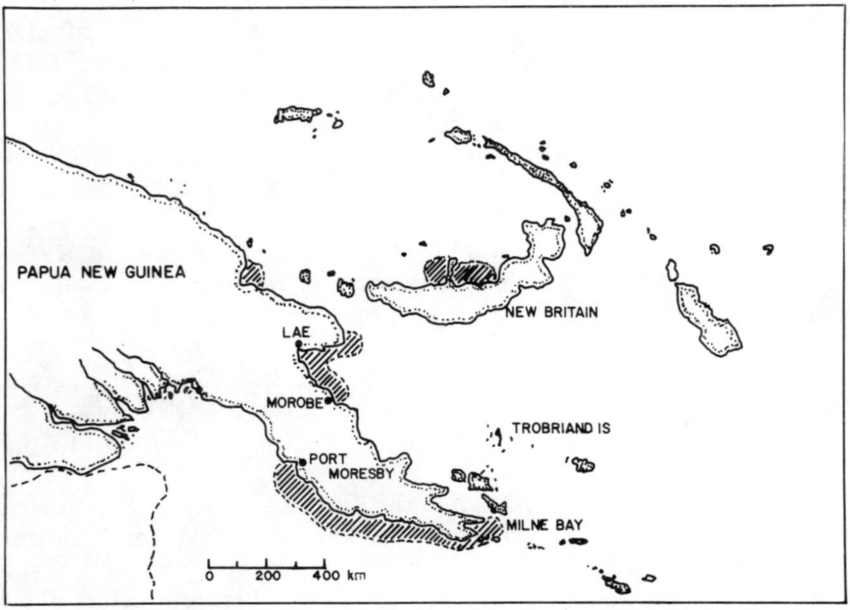

Fig. 1 Chart of Papua New Guinea including the Trobriand Islands and New Britain. The areas of red tides during the 1970's are shown as cross-hatched.

Anecdotal reports of PSP after eating bivalves in the Trobriand Islands, east of PNG and = 300 km north of the easternmost tip of PNG, Milne Bay, go back as far as 1956 (Maclean, this vol.). The native lore is not to eat shellfish when the waters in the lagoons become "phosphorescent" (the dinoflagellate *P. bahamense* var. *compressum* is bioluminescent, similar to *P. bahamense* var. *bahamense* in the Caribbean Sea), implying that the Trobriand Islands have been colonized by *Pyrodinium* for some time. In 1961, red tides during the day, "phosphorescence" during the night (in the same areas as the red tides) and PSP symptoms after eating bivalves, implicated *P. bahamense* var. *compressum* on New Britain Island (≈ 200 km north of the Lae-Huan Gulf area). In April-May 1971, red tides of a chain forming dinoflagellate, tentatively identified as *P. bahamense*, were reported in Port Moresby harbor on the southern coast of PNG and near Lae along the northern coast (Morobe Lagoon area). There were no reports of PSP symptoms.

However in 1972 there were visible red tides and a major outbreak of PSP. The chain-forming *P. bahamense* var. *compressum* was conclusively identified as responsible for red tides in February-May in Port Moresby harbor and southeast along the coastline for approximately 200 km (Maclean 1973, 1975). During October-December of the same year, 1972 a unialgal orange-red tide in Morobe Lagoon (northern coast) was also microscopically identified as *Pyrodinium*. Airplane sightings from 1000 ft altitude indicated additional red tides along the coastline, northwest of Morobe Lagoon, for approximately 60 km (into Lae) (Maclean, pers. comm.).

During Feb-June, 1973, red tides ≈ 100 m wide, verified to be *Pyrodinium*, were observed during weekly air observations, from Port Moresby harbor southeast to Milne Bay on the tip of the PNG peninsula. The red tides were always in shallow waters a few hundred meters from shore inside the barrier reef which parallels the southern shoreline (Maclean, pers..comm.). *P. bahamense* cysts were also found in sediments in the mouth of Port Moresby harbor concurrently (Maclean, pers. comm.). During this same time period, red tides and PSP were observed in waters of west New Britain.

Red tides were also observed in the Port Moresby area in April 1974.

The following interpretation of the data is suggested: Papua New Guinea is in the southern hemisphere. Persistent NW winds during the monsoon season (Jan-May) should produce nearshore coastal upwellings in previously thermally stratified nearshore waters along the northern coast (mirror image of right hand rule in the southern hemisphere) and fronts on the seaward side of the upwelled waters. If *P. bahamense* var. *compressum* were introduced into the seaward side of the frontal region it might be able to grow rapidly. Relaxation of the Northwesterly winds will permit the Warier River surface plume (which empties into the Morobe Lagoon), to produce rapid re-stratification of the nearshore waters. This surface low salinity plume will overlay the more dense offshore surface waters. The reverse flow which is induced will transport these surface waters as a bottom wedge below the pycnocline into the lagoon. Bottom frictional turbulence in the shallows of the lagoon will mix the bottom wedge into the euphotic zone. If this bottom wedge of denser waters from the original coastal front contains dinoflagellates, they will also be turbulently mixed into the euphotic zone, where blooming will ensue. Blooming in the lagoon should be accompanied by phototactic accumulation into red tides along convergence zones of estuarine and plume fronts within the lagoon. Eventually, decreasing nutrient concentrations should stimulate gametogenesis and the production of resting cysts within the dense surface patches and the resting cysts will sink into the shallow sediments.

Toxic dinoflagellates and PSP symptoms were present in New Britain for at least 10 years prior to their first appearance in PNG in 1971 or 1972. The source of *Pyrodinium* for the northern coast of PNG may have been the west coast of New Britain. Dense surface patches of *Pyrodinium* may have been transported from the west coast of New Britain to the upwelling frontal regions by the same sustained northwesterly winds that produced the upwelling fronts off the northern coast of PNG. Unusually persistent currents produced by the northwesterly winds may have resulted in headland fronts off the western tip of New Britain. *Pyrodinium* could have accumulated phototactically into surface patches within the convergence zones of the headland fronts and been transported south and southeast in surface currents to the coastal upwelling fronts along the northern coastline of PNG below Lae. In either case the dinoflagellates would have been transported into nearshore waters at Lae by reverse flow below the pycnocline, as described above.

Surface patches formed at convergence zones of estuarine fronts during tidal periods or at the fronts of low density plumes will drift or be blown out of Morobe Lagoon and other lagoons providing surface lenses for alongshore transport. There are insufficient data to form conclusions as to the mechanisms for the colonization of Milne Bay or whether the southern coast was colonized in steps of emigration from the northern coast.

Sustained southeasterly winds (May-Dec.) should produce nearshore coastal upwellings and offshore upwelling fronts along the southern coast. *Pyrodinium,* mixed into the stratified, seaward sides of the offshore upwelling fronts, might grow to significant concentrations. Relaxation of southeasterly winds, accompanied by geostrophic forcing by surface plumes of runoff waters (occasional rainfall) could result in transport of these dinoflagellates into nearshore waters, again by reverse flow below the pycnocline. The dinoflagellates would be vertically mixed into the euphotic zone of the nearshore waters. Tidal action will exchange the dinoflagellates into Port Moresby Harbor and other lagoons. Once within the Harbor further growth can occur, with phototactic accumulations (red tides and surface lenses) in frontal convergence zones. Nutrients will eventually be utilized to the point where gametogenesis within the dense surface patches may be triggered. As in the case of the northern coast the surface patches formed during tidal periods can drift or be blown out of the Harbor and other lagoons providing surface lenses for alongside transport.

Port Moresby Harbor and Morobe Lagoon now may be annual sources for *P. bahamense* var. *compressum.* It is possible that there exist on the northern or southern coasts of PNG additional protected lagoons in which populations are autochthonous. In the Caribbean Sea lagoons the closely related species, *P. bahamense* var. *bahamense,* was originally found in a small bioluminescent lagoon, "fire-lake" near Nassau, Bahamas (Harvey 1952). It is now autochthonous in Oyster Bay, Jamaica, W.I. and in Bahia Fosforescente and Puerto Mosquito, Puerto Rico (Seliger et al. 1969; 1970; 1971). The var. *bahamense* does not appear to be toxic to fish or shellfish and the author has witnessed natives eating small oysters attached to mangrove roots in Oyster Bay without ill effects.

It is possible that in 1971 or 1972, a fortuitous combination of ocean currents resulted in the surface transport of *Pyrodinium* into the seaward side of upwelling frontal regions on the northern and/or southern coasts of PNG. Port Moresby Harbor certainly, and Morobe Lagoon probably, contain resting cysts which germinate annually under favorable environmental conditions. Therefore a repeat of the unusual wind events of 1971 or 1972 is no longer necessary for the continued annual presence of *Pyrodinium* in these areas. Cysts germinate, vegetative cells grow rapidly, phototactic accumulations (red tides) occur along convergence zones of estuarine or tidal shear or plume fronts. Dense surface patches form which drift out of the embayments and are transported along the shorelines in the direction of the prevailing winds. When nutrients begin to be depleted the dinoflagellates in the dense surface patches will be stimulated to produce sexual gametes which fuse and sink as resting cysts (hypnozygotes) into bottom sediments, to germinate under more favorable conditions.

Sabah and Brunei Darussalam

In 1976, red tides of *P. bahamense* var. *compressum* appeared for the first time on the northwest coast of Borneo, in Sabah (Roy 1977; Wong and ting 1984; Ting and Wong, this vol.) and in Brunei Bay, Brunei Darussalam (Beales 1976; Matdanan Haji Jaafar and Subramaniam 1984). A chart of the entire region is shown in Fig. 2 and the Brunei Bay region in Fig. 3. The approximate extents and locations of the observed red tides are cross-hatched. The red tide observations are abstracted from the above papers and from discussions with M.W.R.N. De Silva, Dept. of Fisheries, Brunei Darussalam and G. Usup, UKM Sabah Campus, Kota Kinabalu, Sabah.

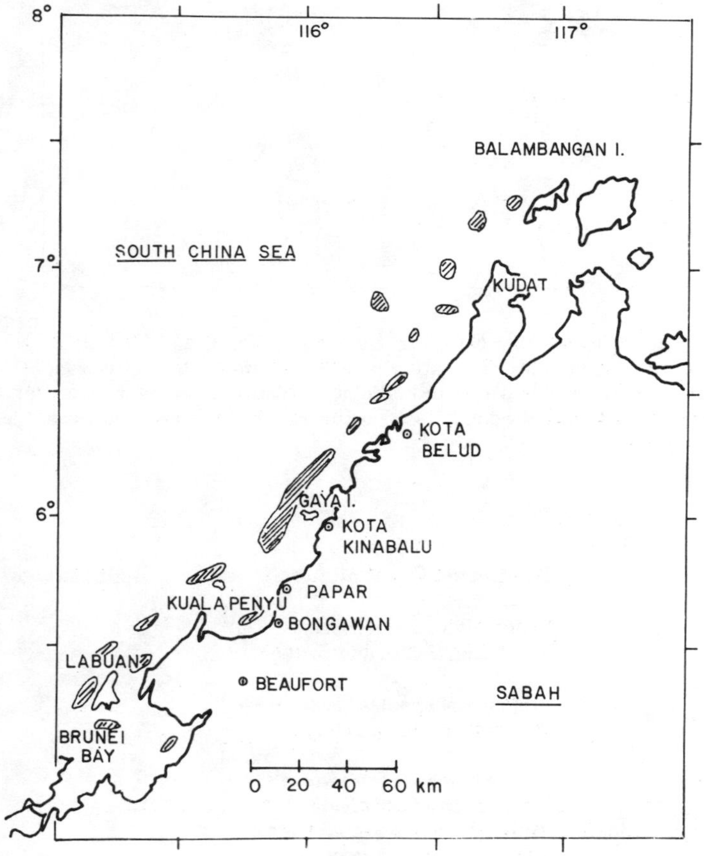

Fig. 2. Chart of the northwest coast of Borneo including Sabah and Brunei Darussalam. The areas where red tides were observed in 1976 and subsequent years are shown as cross-hatched.

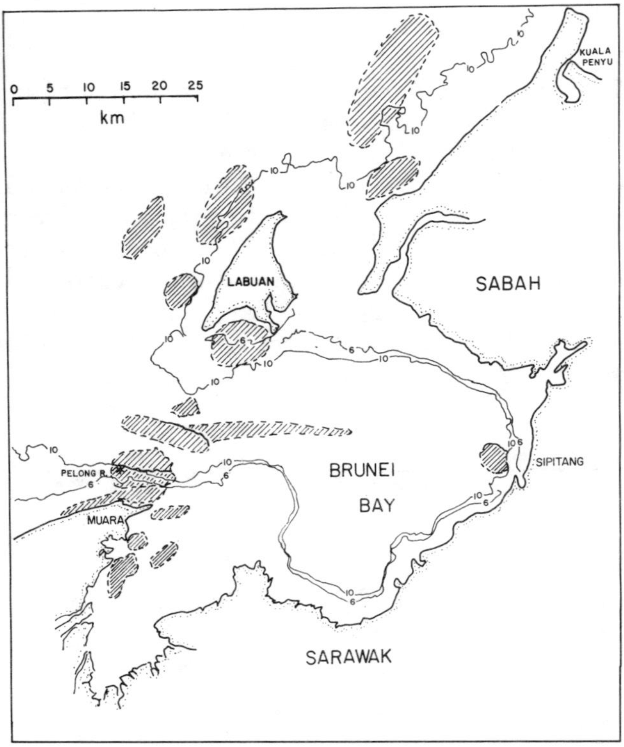

Fig. 3 Chart of Brunei Bay expanded from Fig.2. The 6 and 10 fathom isobaths are shown in thin solid lines. The areas where red tides were observed in 1976 and subsequent years are shown as cross-hatched. The predominant areas for red tides were north & northeast of Pelong Rocks at the mouth of the Bay and the east end of the Bay at Sipitang.

1976	**Northwest Coast of Sabah**	**Brunei Darussalam**
Jan-Feb.	Much higher than average rainfall in Sabah and Brunei	
15 Jan.	PSP symptoms near Kota Kinabalu (5°46'N, 116°E).	
Early Feb.	Dense patches of red water first observed near Kota Kinabalu, ≈ 200 km northwest of and one month prior to the first (11 Mar) sighting in Brunei Bay.	

Continued

	Northwest Coast of Sabah	Brunei Darussalam
11 Mar.		Brick-red patches (≈ 13 km^2) first observed in Brunei Bay, near Pelong Rocks, (≈ 7.4 km north-northeast of Muara Port). PSP symptoms reported.
11-15 Mar.		North-northwesterly winds forced patches southeast to beaches along Brunei Bluff
15 Mar.		Mass PSP symptoms including deaths in Sipitang, at the east end of Brunei Bay (5°5'N, 115°34'E)
16-19 Mar.		Aerial observations. New red tides observed north and northeast of Pelong Rocks. Dispersed within 6 days.
Late Mar.	Peak of red tide outbreak in both Sabah and Brunei, Air observations at ≈ 1,000 feet:	
		Inside Brunei Bay to Sipitang at the east end. There was a distinct colder layer beneath the surface layer.
		Surface patches from inside the Brunei Channel 12 km south of Muara, to the mouth of Brunei Bay.
	From the south and west coasts of Labuan, to Balambangan Island (7°15'N 116°50'E). The alongshore extent was ≈ 300 km.	

Continued

	Northwest Coast of Sabah	Brunei Darussalam
		Brunei appeared to be the southern extent of the red tides. They did not extend into Sarawak on the southwest.
5-12 Apr.		Red tides north and northeast of Pelong Rocks. Dispersed in several days.
14-27 Apr.		Red tides covering large areas inside Brunei Bay and then forming localized dense patches. Patches were observed drifting along the deep water channel of the bay.
4-6 May		New red tides north and northeast of Pelong Rocks. Dispersed by May 6.
Middle of May	Red tides subsided	Red tides subsided

During the period of the red tides (February-April) the weather was hot and sunny, the sea was calm and the winds were light and north-northeasterly. Early May marked the beginning of the southwesterly monsoon season with a shift of the prevailing winds to the southwest. This coincided with the disappearance of the red tides.

1977		
May		Red tide in Brunei Bay. No PSP symptoms reported.
1978-1979		
	No red tides. However PSP symptoms and death near Kota Kinabalu in 1979.	No red tides
1980		
Apr.		Red tide in Brunei Bay.

Continued

	Northwest Coast of Sabah	Brunei Darussalam
1 May		Red tides north and northeast of Pelong Rocks.
1-20 May	PSP and deaths reported southeast of Labuan. Decline of red tides towards end of this time period.	
		Red tides east and south of Muara Port, in streaks in northern part of entrance to the bay, concentrated in a subsurface maximum at 10 meter depth at entrace to bay.
7 July	Decline of red tides	Decline of red tides
Nov.	No red tides.	No red tides
Dec.		Red tides in Brunei Bay.
1981		
Jan-Feb.		Red tides north and northeast of Pelong Rocks.
Feb.-Jul.		Sparse occasional red tides.
July		Red tides north and northeast of Pelong Rocks. Then absent.
1982-1984		
	No red tides. PSP symptoms and deaths along northwest coast.	No red tides
1985		
5 Dec.	Red tides on northwest coast off Kota Kinabalu. Disappeared in 7 days.	

Continued

	Northwest Coast of Sabah	Brunei Darussalam
1986		
9 Dec.	Red tides on northwest coast off Kota Kinabalu. *P. bahamense* var. *compressum* concentrations between 2 x 10^4 and 3.6 x 10^5/l. By 15 December cell concentrations were down to ≈ hundreds/l.	
1987		
	No red tides. PSP symptoms at northen tip of Sabah and near Kota Kinabalu.	No red tides
1988		
11 Jan.	Red tides at Kota Kinabalu.	Red tides in east end of Brunei Bay at Sipitang.
23 Jan.		Subsurface maximum of *P. bahamense* var. *compressum* at 3 m depth north and north-east of Pelong Rocks.
Mar.	Red tides near Kota Kinabalu.	
Jan-May		*P. bahamense* var. *compressum* remained in east end of Brunei Bay at Sipitang at concentrations too small to form visible red tides. In May the organisms bloomed and produced large surface patches in the east end.
1989		
		No red tides.

Northeasterly prevailing winds extend from Janaury-May and southwesterly prevailing winds extend from May-December. The reports indicated that the red tides were not continuous but appeared in phases. Visual sightings of red tides from the air are most intense from 0830h to 1030h and during hot, sunny days with little wind. Colors of red tides varied from orange-red for dense patches in bright sunlight to muddy-brown during overcast days. Sediment streaks near shorelines in bays are sometimes difficult to distinguish from red tides.

ANALYSIS OF DATA: RED TIDES AT THE CONVERGENCE ZONES OF ESTUARINE FRONTS IN BRUNEI BAY; SHORT RANGE DRIFTING OF SURFACE PATCHES IN THE DIRECTION OF THE WIND

It is proposed that the red tides near Pelong Rocks in Brunei Bay are produced in the convergence zones of estuarine fronts. This would also account for their transient nature during the day. In Fig. 2B, 6 and 10 fathom isobaths are drawn inside Brunei Bay. The area north and northeast of Pelong Rocks where red tides are usually observed lies at the southern edge of the mouth of the deep but narrow channel which opens to the South China Sea. This specific area is ideal for the production of estuarine fronts. This bay is both thermally and salinity stratified (see description of reverse flow below the pycnocline, below). The bathymetry is shallow at Pelong Rocks and then just north and northeast drops steeply to the central bay basin. On an ebbing or flooding tide in the thermally stratified bay, frictional turbulence of the bottom, colder waters in the shallow area of Pelong Rocks (just south of the deep channel) will be sufficient to mix vertically the water column. The now denser waters on the shallow southern edge of the deep central channel will sink below the less dense surface layer in the central channel, forming an estuarine front. In bright sunlight the dinoflagellates in these shallow waters will accumulate phototactically along the convergence zone, producing red tides. Dense surface lenses can then drift under the action of the light north and northwesterly winds into the inner bay south of Muara and to form windrows along Brunei Bluff. In May 1989, in the absence of *P. bahamense* var. *compressum* in Brunei Bay, the region north and northeast of Pelong Rocks exhibited streaks and foam lines characteristic of estuarine fronts.

There are two regions in the Bay where stations might be maintained in order to detect the presence of *P. bahamense* var. *compressum* with high sensitivity. These are north-northeast of Pelong Rocks, where estuarine fronts form, and the shallow east end of the Bay opposite Sipitang, where the organisms would be delivered by reverse flow below the pycnocline (see below) and retained because they are in the lee of the mangroves.

Subsurface maximum of dinoflagellate concentrations, reverse flow below the pycnocline

In 1976, Brunei Bay was stratified; the bottom waters below a thermocline at ≈ 4 m were distincly colder (24.5°) than the surface waters (28-29°). In addition the surface layer salinities inside the Bay can vary from 15 to 28 ppt, compared to the bottom salinity around 30.6 ppt, indicating periods of large deliveries of river waters to the Bay. These would produce strongly stratified surface plumes and significant reverse flow below the pycnocline of outside surface waters into the Bay. if the alongshore transport of surface patches of dinoflagellates along the coast of Sabah to the outside of Brunei Bay coincided with a rainfall-produced surface plume, the patches would be overlain by the surface plume and transported into the bay by reverse flow below the pycnocline. During these times it should be possible to determine the depth of the pycnocline and to measure the presence of the dinoflagellates as a subsurface maximum (just below the pycnocline) of chlorophyll (by *in vivo* fluorescence), by microscopic counting of samples from Van Dorn bottle casts or from integrated vertical net tows. In 1980, *P. bahamense* var. *compressum* in Brunei Bay appeared to be concentrated at 10m depth. In 1988, these dinoflagellates were below 3 m but their precise depth distributions were not determined.

If the estuarine fronts occur at night or the area is overcast during the day, positive phototaxis will not be too effective. The sinking waters of the estuarine front will carry the dinoflagellates below the stratified surface waters of the central channel. Since the sinking waters are intermediate in density between the least dense surface and the most dense bottom of the stratified system, they will sink to just below the thermocline and will produce a subsurface maximum of dinoflagellates inside the mouth of the Bay. During periods of high fresh water delivery to the Bay the reverse flow below the Bay pycnocline will deliver the dinoflagellates to the shallow eastern end, as far as Sipitang, where the bottom waters will be turbulently mixed throughout the euphotic zone.

Alongshore transport and upwelling along the northwest coast

It is possible that *P. bahamense* var. *compressum* was autochthonous in one of the small lagoons in Sabah prior to 1976. Candidates might be the lagoons along Kota Kinabalu, Kota Belud of Kudat at the northern tip of Sabah. The lagoon at Kuala Penyu (5°33'N, 115°35'E) closely resembles the bioluminescent bay, Puerto Mosquito in Vieques, Puerto Rico, in size, shape, aspect to the prevailing winds and in mangroves protecting it from wind. The time sequence of the 1976 outbreak of *Pyrodinium* in northwest Borneo was such that the direction of transport was most likely from Kota Kinabalu, ≈ 100 km northeast of Brunei Bay, to the South China Sea outside the Bay during a period of prevailing northeasterly winds. From there a low salinity surface plume extending into the coastal waters provided the energy for the delivery of the surface patches into the Bay by reverse flow below the pycnocline. Kuala Penyu lagoon appears to be too far south for it to have been the source of the 1976 outbreak at Kota Kinabalu.

The northeasterly winds produce upwelling in the nearshore waters along the coast and upwelling frons on the seaward side of the upwelling regions. Possibly sometimes in January 1976, abnormally heavy runoff due to rainfall may have flushed *Pyrodinium* out of its lagoon on the coast of Sabah somewhere north of Kota Kinabalu and into coastal waters. The appearance of PSP symptoms and red tides at Brunei Bay followed those at Kota Kinabalu by approximately one month. There would have been sufficient time for stepwise alongshore transport to both the Kota Kinabalu area and Brunei Bay, followed by delivery into each of these areas by reverse flow below the pycnocline.

The areas of red tides along the coast of Sabah, shown in Fig. 2A, and the red tides in Brunei Bay imply that *Pyrodinium* has now colonized lagoons near Kota Kinabalu and Brunei Bay, confirmed from the recent finding of resting cysts of *P. bahamense* var. *compressum* in sediments in Brunei Bay and in Kuala Penyu. However from the time courses of the dissipations of red tides in Brunei Bay and from the observations of the absence of red tides and PSP symptoms in some of the years following 1976, it does not appear that there is significant annual retention of *Pyrodinium* in Brunei Bay. In the absence of anomalous wind events or if the hypereutrophication that enhanced the alongshore transport were reduced, Brunei Bay may eventually become free of *Pyrodinium*.

Inadvertent Delivery

P. bahamense var. *compressum* could have been inadvertently released into the southern or northern coasts of Papua New Guinea, or the northwest coast of Sabah or Brunei Bay from the water ballasts of ships (possibly oil tankers in the case of northwest Borneo) coming from distant areas where the dinoflagellates were already established and from which transport via ocean surface currents might not be possible. Once inoculated into these waters the organisms might have found ideal colonization conditions: eutrophic waters for rapid vegetative growth; embayments and prevailing winds for retention of the populations; estuarine, tidal shear and/or plume fronts for phototactic accumulations in convergence zones (red tides) and into surface patches where gametogenesis and resting cyst formation can occur when nutrients become exhausted. Such surface patches can be transported to new areas in alongshore surface currents; and shallow bottom sediments in which resting cysts formed and deposited under low-nutrient conditions, can be readily stirred and germinated under favorable environmental conditions.

Conclusions: Stepwise Alongshore Transport

The following listing does not represent an exhaustive survey of red tide events. However it does cover first time occurrences of red tides in new geographical areas.

The first serious outbreak of PSP symptoms caused by *Alexandrium tamarense* in the southern Bay of Fundy, in Canada bordering northeastern Maine, occurred in 1957. In subsequent years shellfish beds in eastern Maine were closed and continue to be closed (Shumway et al. 1988).

During 1971/1972, another El Niño period of worldwide anomalous wind and storm events, red tides of toxic dinoflagellates were apparently transported to and appeared precipitously for the first time in two other geographical regions where they had not been previously observed. These were *A. tamarense* in the southern Gulf of Maine, U.S.A. (Mulligan 1975; Hartwell 1975; Yentsch et al. 1975) as far south as Cape Cod, Massachusetts, (Anderson and Morel, 1979a; 1979b) and *Gymnodinium breve* along the southeast coast of Florida, U.S.A. (Murphy et al. 1975; Steidinger 1975). The appearances of *A. tamarense* and *G. breve* in new areas in 1972 have been attributed to unusual wind events and ocean surface currents in which the organisms were apparently transported for large distances (Mulligan 1975; Murphy et al. 1975). The progression of *A. tamarense* appears to be stepwise (not necessarily all steps in the same year), presumably from the St. Lawrence estuary to the Bay of Fundy, to the east coast of Maine, to Cape Cod, Massachusetts. *G. breve* emigrated from the Tampa Bay-Sanibel Island area on the west coast of Florida, initially southeast to the Straits of Florida and then around the Florida peninsula to the southeastern coast, a total distance of 300-400 km. There are two conditions for successful transport. Not only is it necessary for the wind-driven currents to be persistent in the proper direction and to occur at the times when the dinoflagellates at the source have bloomed and formed into surface patches, but concurrently it is necessary that nutrients be present in the coastal waters along the transport path and be mixed partially into the surface currents in which the dinoflagellate patches are entrained. This permits dinoflagellate growth and positive phototaxis to compensate for dilution out of the jets by water exchange. The combinations of winds and currents that occurred in the Gulf of Maine and around the Florida coasts during the El Niño year of 1972 have very likely occurred in the past without resulting in new colonizations. However, by 1972 nutrient concentrations in coastal waters and potential for reproduction of dinoflagellates along the transport paths may have been sufficient to compensate for dilution losses of transported dinoflagellates.

In the same year that *P. bahamense* var. *compressum* red tides and PSP symptoms were first observed in Sabah and in Brunei-Darussalam in 1976, the Atlantic coast of Spain recorded the first outbreak of *Gymnodinium catenatum* red tides and extensive PSP toxin in their cultured mussel industry. In 1983 *P. bahamense* var. *compressum* red tides and PSP symptoms were first observed in the Philippines (Estudillo and Gonzales 1984).

These six new colonizations (seven counting Papua New Guinea) in geographical areas hundreds of kilometers removed from their apparent sources, correlate with El Niño years of anomalous wind events. This implies that unusual combinations of wind-driven transports were possible. However these wind and current anomalies are not so improbable that they might not have occurred in the past. These red tide dinoflagellates have been shown to produce resting cysts, presumably to insure retention during periods of unfavorable environmental conditions. Therefore once an area is colonized the organisms should become (and have been reported to become) permanent or at least annual seasonal residents. Why then, have so many different geographical areas been only recently colonized, and why are there no anecdotal records of PSP due to possible earlier transports?

One possible suggestion is that higher nutrients along parts of transport paths have made it more probable for dinoflagellates to emigrate successfully via alongshore transport and to colonize new areas. Embayments in which dinoflagellate red tides have been observed for many years may have been colonized prior to anthropogenic eutrophication. Some steady state must have been reached where further stepwise emigration, along all possible surface current pathways, was nutrient-limited. Coastal hypereutrophication due to deforestation and population growth has varied in degree and in time in different geographical areas. Recent occurrences of red tides in new geographical areas may thus be indicative of relatively recent coastal nutrient increases somewhere along the transport paths. Areas recently colonized are apparently serving as sources for further emigration and colonization.

References

Anderson, D.M. and F.M.M. Morel 1979a. The seeding of red tide blooms by the germination of benthic *Gonyaulax tamarensis* hypnocysts. Est. Coast. Mar. Sci. 8: 279-293.

Anderson, D.M. and F.M.M. Morel. 1979b. Toxic dinoflagellate blooms in the Cape Cod region of Massachusetts, p. 145-150. *In* D.L. Taylor and H.H. Seliger (eds.) Toxic dinoflagellate blooms, Elsevier, New York.

Beales, R.W. 1976. A red tide in Brunei's coastal waters Brunei Mus. J. 3(4): 167-182.

Dale, B. 1983. Dinoflagellate resting cysts: 'benthic plankton', p. 69-136. *In* G.A. Fryxell (ed.) Survival strategies of the algae. Cambridge University Press, New York.

Dale, B., C.M. Yentsch and J.W. Hurst. 1978. Toxicity in resting cysts of the red-tide dinoflagellate *Gonyaulax excavata* from deeper water sediments off the Maine coast, U.S.A. Science 201: 1223-1225.

Hartwell, A.D. 1975. Hydrographic factors affecting the distribution and movement of toxic dinoflagellates in the western Gulf of Maine, p. 47-68. *In* V.R. LoCicero (ed.) Proc. 1st Internat. Conf. Toxic dinoflagellate blooms. Mass. Sci. Technol. Found. Wakefield, Mass.

Harvey, E.N. 1952. Bioluminescence. Academic Press, New York. 649 p.

Jaafar, Matdanan Haji and S. Subramaniam. 1984. Occurrences of red tide in Brunei Darussalam and methods of monitoring and surveillance, p. 17-24. *In* A.W. White, M. Anraku and K.K. Hooi (eds.) Toxic red tides and shellfish toxicity in southeast Asia, Southeast Asian Fisheries Development Center and International Development Research Centre, Singapore.

Maclean, J.L. 1973. Red tide and paralytic shellfish poisoning in Papua New Guinea. Papua New Guinea Agric. J. 24(4): 131-138.

Maclean, J.L. 1975. Paralytic shellfish poison in various bivalves, Port Moresby 1973. Pac. Sci 29(4): 349-352.

Maclean, J.L. 1979. Indo-Pacific red tides, p. 173-178. *In* D.L. Taylor and H.H. Seliger (eds.) Toxic dinoflagellate blooms, Elsevier, New York.

Maclean, J.L. 1984. Indo-Pacific toxic red tide occurrences, 1972-1984, p. 92-104. *In* A. W. White, M. Anraku and K.K. Hooi (eds.) Toxic red tides and shellfish toxicity in southeast Asia. Southeast Asian Fisheries Development Center and International Development Research Centre, Singapore.

Mulligan, H.F. 1975. Oceanographic factors associated with New England red tide blooms, p. 23-40. *In* V.R. LoCicero (ed.) Proc. 1st Internat. Conf. Toxic dinoflagellate blooms. Mass, Sci. Technol. Found. Wakefield, Mass.

Murphy, E.B., K.A. Steidinger, B.S. Roberts, J. Williams and J.W. Jolley, Jr. 1975. An explanation for the Florida east coast *Gymnodinium breve* red tide, November, 1972. Limnol. Oceanogr. 20: 481-486.

Roberts, B.S. 1979. Occurrence of *Gymnodinium breve* red tides along the west and east coasts of Florida during 1976 and 1977 p. 199-202. *In* D.L. Taylor and H. H. Seliger (eds.) Toxic dinoflagellate blooms, Elsevier, New York.

Roy, R.N. 1977. Red tide and outbreak of paralytic shellfish poisoning in Sabah. Med. J.Malays. 31(3): 247-251.

Sasner, J.J. Jr., M. Ikawa and B.E. Barrett 1975. The 1972 red tide in New Hampshire p. 517-522. *In* V.R. LoCicero (ed.) Proc. 1st Internat. Conf. Toxic dinoflagellate blooms. Mass. Sci. Technol. Found. Wakefield, Mass.

Seliger, H.H., W.G. Fastie and W.D. McElroy. 1969. Towable photometer for rapid area mapping of concentrations of bioluminescent marine dinoflagellates. Limnol. Oceanogr. 14: 806-813.

Seliger, H.H., J.H. Carpenter, M. Loftus and W.D. McElroy. 1970. Mechanisms for the accumulation of high concentrations of dinoflagellates in a bioluminescent bay. Limnol. Oceanog. 15: 234-245.

Seliger H.H., J.H. Carpenter, M. Loftus, W.H. Biggley and W.D. McElroy. 1971. Bioluminescence and phytoplankton succesions in Bahia Fosforescente, Puerto Rico, Limnol. Oceanogr. 16: 608-622.

Seliger, H.H., M.E. Loftus and D.V. SubbaRao. 1975. Dinoflagellate accumulations in Chesapeake Bay, p. 181-205. *In* V.R. LoCicero (ed.) Proc. 1st Internat. Conf. Toxic dinoflagellate blooms. Mass, Sci. Technol. Found. Wakefield, Mass.

Shumway, S.E., S. Sherman-Caswell and J.W. Hurst. 1988. Paralytic shellfish poisoning in Maine: Monitoring a monster J. Shellfish Res. 7(4): 643-652.

Steidinger, K.A. 1975. Basic factors influencing red tides, p. 153-162. *In* V.R. LoCicero (ed.) Proc. 1st Internat. Conf. Toxic dinoflagellate blooms. Mass. Sci. Technol. Found. Wakefield, Mass.

Steidinger, K.A. and K.D. Haddad. 1981. Biologic and hydrographic aspects of red tides. Biosci. 31: 814-819.

Toxins in *Pyrodinium bahamense* var. *compressum* and Infested Marine Organisms

YASUKATSU OSHIMA
Faculty of Agriculture
Tohoku University
1-1 Tsutsumidori Amamiyamachi
Sendai 981, Japan

OSHIMA, Y. 1989. Toxins in *Pyrodinium bahamense* var. *compressum* and infested marine organisms, p. 73-79. *In* G.M. Hallegraeff and J.L. Maclean (eds.) Biology, epidemiology and management of *Pyrodinium* red tides. ICLARM Conference Proceedings 21, 286 p. Fisheries Department, Ministry of Development, Brunei Darussalam, and International Center for Living Aquatic Resources Management, Manila, Philippines.

Abstract

The toxin profiles of Philippines green mussels *(Perna viridis)* and Borneo planktivorous fish (*Sardinella* sp.) contaminated by the dinoflagellate *Pyrodinium bahamense* var. *compressum,* were characterized as (neoSTX), STX, GTX5 and dcSTX. These results are comparable with previous studies on Palauan dinoflagellates and confirm the absence of production of 11-hydrosulfate toxins by this PSP dinoflagellate. High toxin levels in *Sardinella* may explain the high incidence of human poisonings from *Pyrodinium* red tides.

Introduction

It is now well recognized that most paralytic shellfish poisonings occurring in the tropical Pacific regions are related to blooms of the dinoflagellate *Pyrodinium bahamense* var. *compressum*. From the toxicological point of view, the presence of potent neurotoxins in this organism was first shown by Maclean (1975) using the mouse bioassay on specimens from Papua New Guinea. However, the chemical nature of toxins was not clarified until our study on Palauan specimens (Harada et al. 1982a). Analysis of toxins in *P. bahamense* var. *compressum* and infested shellfish by means of conventional purification procedures with an ion exchange column and successive thin layer chromatographic analysis revealed the presence of neosaxitoxin (neoSTX), saxitoxin (STX), gonyautoxin 5, 6 (GTX5, GTX6) and an unidentified toxin tentatively coded as PBT1 (Harada et al. 1982a). The structures of GTX5 and GTX6, which had not been elucidated at that time, were shown to be N-sulfocarbamoylsaxitoxin and N-sulfocarbamoylneosaxitoxin, respectively (Harada et al. 1982b), simultaneously with other investigators who worked on the same toxins from *Alexandrium (Protogonyaulax)* spp. PBT1 later proved to be decarbamoylsaxitoxin by spectroscopic analysis and other methods, and was the first recognition of its occurrence in the natural environment (Harada et al. 1983).

So far 18 saxitoxin analogs (Fig. 1) are known to occur naturally, isolated mainly from samples contaminated with *Alexandrium* spp. It

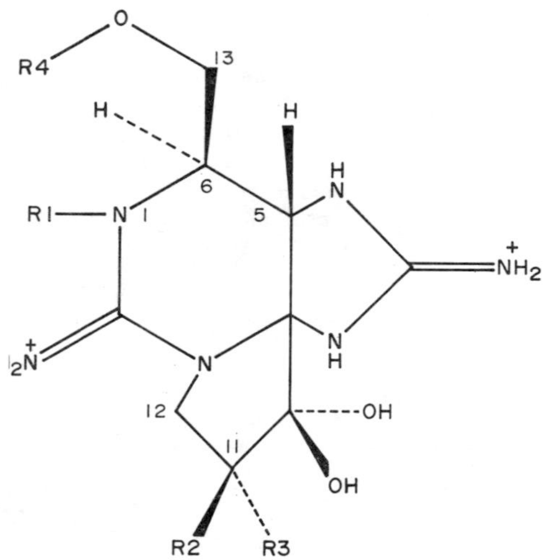

			Carbamate toxins		N-sulfo-carbamoyl toxins		Decarbamoyl toxins	
R1	**R2**	**R3**	**R4:CONH$_2$**		**R4:CONHSO$_3^-$**		**R4:H**	
H	H	H	STX	[2045]	GTX5	[350]	dcSTX	[1220]
OH	H	H	neoSTX	[1038]	GTX6	[180]	dcneoSTX	-
OH	H	OSO$_3^-$	GTX1	[1638]	C3	[18]	dcGTX1	-
H	H	OSO$_3^-$	GTX2	[793]	C1	[16]	dcGTX2	[530]
H	OSO$_3^-$	H	GTX3	[2234]	C2	[430]	dcGTX3	[990]
OH	OSO$_3^-$	H	GTX4	[673]	C4	[57]	dcGTX4	-

[ ] = toxicity in MU/μmole.

Fig. 1. Structures and toxicities of paralytic shellfish toxins.

has been noticed that from great variation of toxin profiles exists among isolates of *Alexandrium* spp. For the determination of toxin profiles with a small amount of sample low toxin concentration, sensitive and specific high performance liquid chromatographic (HPLC)-fluorometric methods have been developed by several researchers (Oshima et al. 1984; Sullivan et al. 1985; Nagashima et al. 1987; Oshima et al. 1988). Application of one of the systems to mussels from the Philippines and a rock oyster from the Solomon Islands showed the occurrence of the same toxins as those observed in Palauan *Pyrodinium* (Oshima et al. 1987).

This paper deals with toxin profiles of samples from Borneo and the Philippines which were investigated with an improved HPLC method.

Materials and Methods

Sardinella sp. imported from East Malaysia to Brunei Darussalam on December 1987 (De Silva et al., this vol.) and kept frozen at -20°C were used for toxin analysis. The extract of the intestinal contents of some of these fish, prepared immediately after collection at the Department of Fisheries, Brunei Darussalam, and an extract of *Perna viridis* from the Philippines prepared according to the standard mouse bioassay (Williams 1984) by Ms. Maria Mendigo, College of Medicine, University of the Philippines, were used for chemical analysis.

Three specimens of *Sardinella* sp. were individually homogenized in 5 volumes of 0.1 N acetic acid. The homogenate in a tight-capped vial was heated in a boiling water bath for 5 min. and centrifuged at 3,000 rpm for 15 min. Two fish specimens were used for testing the anatomical distribution of toxins. Fish were dissected into 7 tissues - head, gills, intestine, contents of intestine, other visceral parts, abdominal muscle, and dorsal muscle with bones - and subjected to toxin extraction as described above. The fish and shellfish extracts were analyzed by HPLC after passing through a Sep-Pak C18 cartridge column (Waters) which was previously equilibrated with water, and then through an ultrafiltration filter (Millipore, UFC3TGC00) by centrifugation.

Details of the analytical conditions were described in a previous report (Oshima et al. 1989). Briefly, toxins were separated on a C8 bonded silica column (Develosil C8) by isocratic elution with three different mobile phases: (a) tetrabutylammonium in acetate buffer for analysis of C1-C4; (b) 1-heptanesulfonate in ammonium phosphate buffer for gonyautoxins 1-6 (GTX1-6); and (c) mobile phase b plus acetonitrile for the saxitoxin (STX), neosaxitoxin (neoSTX) group.

Toxins in the eluate from the column were continuously oxidized by heating with periodate at 65°C, and the resultant fluorescent compounds were detected by a fluoromonitor at 390 nm with excitation wavelength 330 nm, after acidifying the reaction mixture with acetic acid.

Results and Discussion

For the extraction of toxins, acetic acid was used instead of hydrochloric acid of the standard method, because it is easier to control pH at 3-4 for small samples to prevent hydrolysis of N-sulfocarbamoyl toxins. Typical chromatograms are shown in Fig. 2 and the results are summarized in Table 1, together with those of *P. bahamense* var. *compressum* from Palau. Toxin profiles of the fish and shellfish from Borneo and Philippines were rather simple, composed of STX, neoSTX, dcSTX and GTX5, except for the mussels

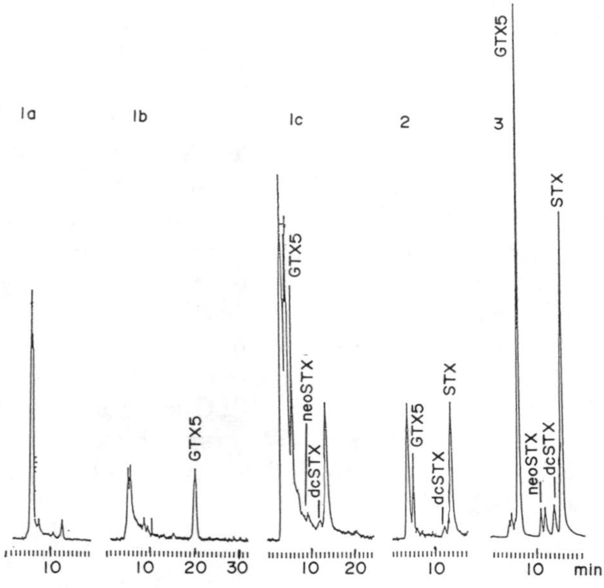

Fig. 2. HPLC-Chromatograms of toxins in the tropical organism. (1) *Sardinella* sp. from Borneo (a-c indicate the mobile phases denoted in the text), (2) *Perna viridis* from Philippines (mobile phase c), and (3) *Pyrodinium bahamense* var. *compressum* from Palau (mobile phase c).

Table 1. Paralytic shellfish toxin composition of some marine organisms from the tropical Pacific.

No. Species (origin)	Organ	Weight (g)	Toxin composition STX	neoSTX	dcSTX	GTX5	Toxin content (nmole/g)	Toxicity[a] (MU/g)	Total toxicity (MU)
1. *Sardinella* sp. (Borneo)	whole	5.49	51.8	29.6	4.0	14.6	9.1	33	181
2. *Sardinella* sp. (Borneo)	whole	6.03	45.7	35.8	2.9	15.7	29.5	41	248
3. *Sardinella* sp. (Borneo)	whole	4.84	34.8	51.0	1.6	12.7	71.3	93	451
4. *Sardinella* sp. (Borneo)	Intestinal contents	0.24	20.5	70.8	3.7	5.0	91.3	73	18
	Intestine	0.25	20.3	68.9	2.0	8.8	59.2	107	27
	Viscera	0.31	24.3	65.3	4.0	6.4	55.8	70	22
	Head	0.88	51.8	28.4	9.2	10.6	16.0	24	21
	Gills	0.22	47.6	28.6	9.5	14.3	38.2	55	12
	Abdominal muscle	0.57	36.1	48.3	5.3	10.3	46.2	62	35
	Dorsal muscle[b]	3.19	24.1	64.4	3.9	7.6	59.1	73	233
	Total	5.66	-	-	-	-	-	65	368
5. *Sardinella* sp. (Borneo)	Intestinal contents	0.05	63.6	29.4	2.0	4.9	80.8	133	7
	Intestine	0.04	63.5	23.3	3.9	9.3	85.0	138	6
	Viscera	0.21	63.9	20.3	4.1	11.5	75.5	122	26
	Head	0.71	63.3	23.6	3.7	9.5	19.8	32	23
	Gills	0.18	81.3	ND	6.1	12.5	21.4	38	7
	Abdominal muscle	0.50	67.6	16.2	5.8	10.4	50.5	122	61
	Dorsal muscle[b]	4.85	48.4	41.6	3.8	6.3	18.8	27	133
	Total	6.54	-	-	-	-	-	40	263
6. *Sardinella* sp. (Borneo)	Intestinal contents[c]	-	33.2	65.0	1.8	ND	-	-	-
7. *Perna viridis* (Philippines)	whole	-	72.4	ND	9.2	18.5	2.6	4	-
8. *Pyrodinium bahamense* var. *compressum*[d] (Palau)		-	28.2	15.3	3.9	47.1	-	-	-

[a]Based on HPLC analysis.
[b]Including bones.
[c]Extract prepared at the Department of Fisheries, Brunei Darussalam.
[d]Containing also 5.5% GTX6.
ND - not detected.

from the Philippines in which neoSTX was under the detection level of HPLC. GTX6 (found in *Pyrodinium* from Palau) and toxins with 11-hydroxysulfate moiety were not detected in any of the samples.

The toxin level of *Sardinella* sp., as whole body, was very high, averaging 54.5 ± 24.8 MU/g which is approximately equivalent to 1,100 µg STX/100 g. Total toxicity per fish was 301 ± 106 MU. The high toxicity levels of the fish observed in this study may well explain the involvement of paralytic shellfish toxins in the human fish poisoning cases which occurred in Sabah at the time of *Pyrodinium* blooms. Toxins were detected in all the fish tissues (Table 1) in contrast to their presence in the intestinal contents only, immediately after catch (M.R.W.N. De Silva, pers. comm.; and De Silva et al., this vol.). The toxins might have migrated into other tissues during storage, since the fish samples were rather dry indicating poor preservation. The higher toxicity in the tissues surrounding the intestine may support the hypothesis. The lower concentration of neoSTX in the gills and head may be due to degradation of the toxin by exposure to air because it is the most labile to oxidation among the toxins found in the fish. The reason for not detecting neoSTX in low toxic *P. viridis* from the Philippines (Sample 7) might also be due to instability of the toxin, as well as poorer response of the HPLC system to neoSTX. The absence of GTX5 in the old *Sardinella* extract (Sample 6) can be explained by conversion of GTX5 to STX by hydrolysis during storage, since the extract showed pH less than 2.

The reported poisonings due to fish consumption in Borneo (Maclean 1979) are unique compared to the toxic events occurring in the northern part of the world. It is easy to speculate that high density of the causative organism and dietary preference of the fish for *P. bahamense* var. *compressum* may be the reason for the high toxin content in *Sardinella* sp. However, more detailed investigation on the physiology of *Sardinella* sp. is necessary to clarify the mechanism of fish poisoning, such as resistance of the fish to toxins or ability to accumulate toxins in organs other than the intestinal tract. From the view of public health, involvement of fish as the vector of paralytic shellfish toxins to humans raises a serious problem. Since the fish might migrate far from the bloom area, surveillance in a wider area is necessary.

Complete absence of 11-hydroxysulfate toxins, such as GTX1-GTX4 and C1-C4, in this study might be a reflection of the lack of ability to produce this toxin group by *P. bahamense* var. *compressum* in Borneo and the Philippines, similar to the Palauan specimens. Thus, *P. bahamense* var. *compressum,* including Guatemala specimens (Rosales-Loessener et al. 1989), appears to have less diversity in toxin profile than *Alexandrium* species.

Acknowledgements

The author wishes to express his sincere thanks to G. Hallegraeff (CSIRO, Australia) and J.L. Maclean (ICLARM, Philippines) for giving him the chance to visit Brunei Darussalam and also to Prof. T. Nemoto (Ocean Research Institute, Tokyo University, the chairman of WESTPAC/IOC) for financial support. Thanks are also due to Ms. Siti Amin Mahali (Department of Fisheries, Brunei Darussalam) and Ms. Kiyoko Sugino (Tohoku University) for technical assistance.

References

Harada, T., Y. Oshima, H. Kamiya and T. Yasumoto. 1982. Confirmation of paralytic shellfish toxins in the dinoflagellate *Pyrodinium bahamense* var. *compressa* and bivalves in Palau. Nippon Suisan Gakkaishi 48: 821-825.

Harada, T., Y. Oshima and T. Yasumoto. 1982. Structures of two paralytic shellfish toxins, gonyautoxins V and VI, isolated from a tropical dinoflagellate, *Pyrodinium bahamense* var. *compressa*. Agric. Biol. Chem. 46: 1861-1864.

Harada, T., Y. Oshima and T. Yasumoto. 1983. Natural occurrence of decarbamoylsaxitoxin in tropical dinoflagellate and bivalves. Agric. Biol. Chem. 47: 191-193.

Maclean, J.L. 1975. Red tide in the Morobe district of Papua New Guinea. Pac. Sci. 29: 7-13.

Maclean, J.L. 1979. Indo-Pacific red tides, 1979, p. 173-178. *In* D.L. Taylor and H.H. Seliger (eds.) Toxic dinoflagellate blooms. Elsevier, New York.

Nagashima, Y. J. Maruyuma, T. Noguchi and K. Hashimoto. 1987. Analysis of paralytic shellfish poison and tetrodotoxin by ion-pairing high performance liquid chromatography. Nippon Suisan Gakkaishi 53: 819-823.

Oshima, Y., M. Machida, K. Sasaki, Y. Tamaoki and T. Yasumoto. 1984. Liquid chromatographic-fluorometric analysis of paralytic shellfish toxins. Agric. Biol. Chem. 48: 1707-1711.

Oshima, Y., T. Yasumoto, G. Hallegraeff and S. Blackburn. 1987. Paralytic shellfish toxins and causative organisms in the tropical Pacific and Tasmanian waters, p. 423-428. *In* P. Gopalakrishnakone and C.T. Tan (eds.) Progress in venom and toxin research. National University of Singapore, Singapore.

Oshima, Y., K. Sugino and T. Yasumoto. 1988. Latest advances in HPLC analysis of paralytic shellfish toxins, p. 319-326. *In* S. Natori, K. Hashimoto and Y. Ueno (eds.) Mycotoxins and phycotoxins '88. Elsevier, New York.

Rosales-Loessener, F., E.D. Porras and M.W. Dix. 1989. Toxic shellfish poisoning in Guatemala, p. 113-116. *In* T. Okaichi, D.M. Anderson and T. Nemoto (eds.) Red tides, biology, environmental science and toxicology. Elsevier, New York.

Sullivan, J.J., J. Jonas-Davis and L.L. Kentala. 1985. The determination of PSP toxins by HPLC and autoanalyzer, p. 275-280. *In* D.M. Anderson, A.W. White and D.G. Baden (eds.) Toxic dinoflagellates. Elsevier, New York.

Williams, S., editor. 1984. Paralytic shellfish poison, p. 344-345. *In* Official methods of analysis. 14th ed. Association of Official Analytical Chemists, Arlington.

Cysts as Factors in *Pyrodinium bahamense* Ecology

DONALD M. ANDERSON
Biology Department
Woods Hole Oceanographic Institution
Woods Hole, MA 02543 USA

ANDERSON, D.M. 1989. Cysts as factors in *Pyrodinium bahamense* ecology, p. 81-88. *In* G.M. Hallegraeff and J.L. Maclean (eds.) Biology, epidemiology and management of *Pyrodinium* red tides. ICLARM Conference Proceedings 21, 286 p. Fisheries Department, Ministry of Development, Brunei Darussalam, and International Center for Living Aquatic Resources Management, Manila, Philippines.

Abstract

Pyrodinium bahamense, like many red tide-forming dinoflagellates, includes a dormant cyst stage in its life history. Given the important role of such resting stages in bloom initiation, species dispersal, genetic recombination, survival through environmental or nutritional stresses, and as vectors of toxicity, there is a clear need to study the distribution, abundance, and general physiological ecology of *P. bahamense* cysts. This paper summarizes what little is known about the cysts of this important species and suggests a variety of research programs that would add much to our understanding of the spatial and temporal dynamics of *P. bahamense* red tides.

Introduction

The resting cyst is an important factor in the ecology of many neritic dinoflagellates. These highly resistant cells can survive in the sediment for extended periods of time and then germinate to release vegetative cells that serve as an inoculum to initiate blooms. Cysts are considered important with respect to species dispersal, bloom timing, bloom location, survival through adverse conditions, and, for toxic species, as sources of toxin (Wall 1971; Anderson 1984). Twenty years ago, Wall and Dale (1969) demonstrated that a microfossil called *Hemicystodinium zoharyi* (Rossignol 1961) was in fact the resting cyst of *Pyrodinium bahamense.* Aside from the preliminary germination experiments conducted by Wall and Dale (1969), all other published work on the living form of this cyst has been taxonomic in nature (e.g., Matsuoka et al. 1989). The objective of this paper is to describe the information that has been obtained for other cyst-forming dinoflagellates and to discuss how such data can be useful in studies of *P. bahamense* bloom dynamics and distribution.

Life Histories

The dominant reproductive mode in dinoflagellates is asexual fission (reviewed in Pfiester and Anderson (1987)). Those species that form true resting cysts are also capable of sexual reproduction, forming gametes that fuse into a swimming zygote (planozygote) which, in most cases, transforms into a nonmotile, thick-walled resting cyst (Beam and Himes 1980). The morphological characteristics of these different life history stages and the duration of the transitions between stages and of the stages themselves differ between species. Examples of complete dinoflagellate life history descriptions are those of *Gymnodinium pseudopalustre* (von Stosch 1973), *Peridinium cinctum* (Pfiester 1975), and *Gyrodinium uncatenum* (Tyler et al. 1982).

Induction of sexuality in laboratory cultures is most often accomplished by nutrient limitation (e.g., Pfiester 1975; Turpin et al. 1978; Anderson et al. 1984; Anderson et al. 1985), but cyst formation has also been observed in nutrient-replete media (Yoshimatsu 1981; Watanabe et al. 1982). One of the problems in studying dinoflagellate sexuality in culture is that not all cultures will produce cysts, even when many potential mating types are crossed (Anderson et al. 1984). This may relate to unknown culture stresses or to genetic changes in species maintained in culture for extended intervals.

None of the information described above is known for *P. bahamense.* Research is needed to describe the details of the *P. bahamense* life cycle, with emphasis on: mating types; morphology of gametes and planozygotes; and mechanisms of sexual induction. The ability to recognize these life cycle stages in natural samples will provide important information on the stage of the bloom (i.e., initiation, development or decline). Likewise, studies that work out the steps needed to produce cysts in laboratory cultures may be necessary for subsequent efforts to understand the physiological or biochemical characteristics of the cysts of this species.

Dormancy and Quiescence

A newly formed cyst generally has a "maturation" interval during which it cannot germinate, even when growth conditions are optimal (Pfiester and Anderson 1987). Termed "dormancy", this stage should be distinguished from "quiescence", which refers to the subsequent interval when maturation is complete, but the resting state continues due to the absence of suitable growth conditions. Domancy is under internal (endogenous) control, whereas quiescence is maintained by external (exogenous) factors.

The duration of dormancy varies significantly between species. For some, it can last several months (e.g., *Alexandrium tamarense (Protogonyaulax tamarensis)* Anderson 1980), a few weeks (*Scrippsiella trochoidea;* Binder and Anderson 1987), or a few days (*Gymnodinium catenatum;* Blackburn et al. 1989). This maturation time interval has major implications on the bloom dynamics of a species. For example, the relatively long dormancy requirements of *A. tamarense* mean that cysts deposited after a bloom cannot germinate for several months, sometimes leading to two distinct bloom events in the spring and fall, respectively, in temperate waters (Anderson and Morel 1979). In contrast, the much shorter dormancy of *S. trochoidea* allows that species to cycle rapidly between the sediments and the water column, resulting in nearly continuous presence of motile cells when temperatures are favorable for growth (Binder and Anderson 1987).

Wall and Dale (1969) described a series of germination experiments using *P. bahamense* cysts that suggest the existence of a dormancy interval. Sediment collected in February contained cysts that did not germinate in mid-March under conditions that did result in germination several weeks later. Much more detailed experiments are required to obtain further information.

When dormancy is complete and the cyst is capable of germination, ɖuiescence can be maintained by a variety of external factors. Three of the most important are sub- or supra-optimal temperatures, anoxia and inadequate light. In temperate waters, where cysts can experience dramatic seasonal changes in bottom temperatures, data now being collected show that many species have a temperature range outside which germination is inhibited (Pfiester and Anderson 1987). For example, cysts of *A. tamarense* from Cape Cod (USA) germinate between 6 and 20°C, whereas cyst of *Gyrodinium uncatenum* from the same waters only germinate between 15 and 21ÁC (D.M. Anderson, unpub. data). It is thus not surprising that the former species first appears in the spring and the latter in the summer/early fall in that region.

Similar data, if obtained for *P. bahamense,* would provide

insights into the role of temperature in bloom initiation. Such information would be especially interesting given the relatively narrow range of temperature fluctuations in tropical waters. A hint of low temperature inhibition of *P. bahamense* cyst germination is found in the study by Wall and Dale (1969) which indicated that there was no cyst germination during 6 months of storage at 16°C. A more complete study is clearly warranted.

A study by Anderson et al. (1987) demonstrated that the cysts of several different dinoflagellates could not germinate in anoxic sediments. Another study (Anderson et al. 1982) showed that most cysts in core samples are not located at the sediment surface but are buried by bioturbation and sedimentation and are thus below the oxidized layer at the surface where germination is possible. *P. bahamense* may be no exception and special experiments on this species need not be a high priority. On the other hand, vertical cyst distribution needs to be studied to provide the background to evaluate the potential impact of anoxia on *P. bahamense* bloom initiation.

Another factor that affects cyst quiescence is light, which would be absent or low due to cyst deposition in deeper waters or burial within the sediments. A wide range of responses to light have been observed between species. Some will not germinate in the dark, others require only a brief pulse of light to initiate germination, while most can germinate in the dark but at rates that are slower than those observed with illumination (Binder and Anderson 1986; Anderson et al. 1987). Here again, nothing is known of the light requirements of *P. bahamense* cysts.

Cysts that have matured through their dormancy phase can re-enter the resting state and be unresponsive to favorable growth conditions (Anderson and Keafer 1987). In higher plants, this process is termed secondary dormancy, with the external environment (e.g., seasonal temperature changes) causing the transition. The process discovered in cysts of *A. tamarense* is similar, except that the evidence suggests that the re-induction of true dormancy is under internal control - that there is an endogenous annual clock that determines germination potential (Anderson and Keafer 1987). An endogenous rhythm if present in *P. bahamense* cysts, could explain the timing of blooms given the relatively small temperature ranges of tropical waters.

Cyst Distributions

Two types of cyst distribution studies have been conducted - qualitative and quantitative. The former typically determine the presence or absence of the cyst of a species in a series of samples along a coastline (e.g., Anderson et al. 1982). These studies are valuable in describing the general geographic distribution of a species and can even indicate areas where toxicity might be an unrecognized problem. For example, the detection of cysts of *A. tamarense* in areas of Connecticut (USA) with no history of shellfish toxicity suggested a re-evaluation of monitoring station locations and eventually led to the detection of toxicity in shellfish in areas that were previously not monitored (Anderson et al. 1982). Although inherently quite valuable and easy to obtain, presence versus absence cyst data should be interpreted with caution. The presence of cysts of a species of interest does suggest the potential for bloom initiation, but the converse is not true. The absence of detectable cysts of a species does not mean an area is free from the threat of toxic blooms. An excellent example of this point emerged during this *Pyrodinium* workshop, where multiple sediment samples collected in Brunei Bay were found to contain very few living dinoflagellate cysts. Since the cores were taken in areas where large *P. bahamense* blooms had occurred in recent years, the paucity of cysts clearly reflects factors such as cyst resuspension and transport and not the continual absence of the species from the plankton.

Quantitative cyst mapping is a much more time-consuming, tedious process that should only be attempted if specific goals justify the effort. Such data provide countour plots of cyst distributions that can reveal important patterns. One example that may be especially relevant to the *P. bahamense* problem in the Indo-West pacific region is a study of *A. tamarense* cysts in an estuarine system on Cape Cod (Anderson et al. 1982). That mapping effort showed how cysts were numerous within one lagoon or embayment but undetectable in adjacent waters. This suggested a "point source" for the species - a location where blooms can originate before being transported elsewhere. Other cyst surveys (Tyler et al. 1982; White and Lewis 1982; Anderson and Keafer 1985) provide examples of how widespread cyst distributions can be depicted with quantitative mapping techniques. Consistent with our general lack of knowledge on all aspects of *P. bahamense* biology and ecology, there have been no cyst distributional studies of any type. In addition to the motivations described above, the tantalizing relationship that is often suggested between *P. bahamense* blooms and mangrove systems could be strengthened by cyst surveys that reveal an abundance of cysts in nearshore waters, possibly even in localized "point sources" near mangrove-dominated areas.

The distributions depicted by mapping surveys should be interpreted with caution. High cyst concentrations in certain locations may well indicate important seedbeds for bloom initiation. However, the cysts may be accumulating in areas where light levels or temperatures are unfavorable for germination, or where germination might produce motile cells that are consistently carried away from shellfish or fish resources. One means to assess the magnitude of the inoculum provided by cysts at a location is to conduct a quantitative survey of the site, and then to monitor the extent of germination. This can be accomplished by monitoring the chlorophyll fluorescence that is a sign of impending germination for some species (Yentsch et al. 1980; Anderson and Keafer 1985). The highly resistant cyst wall of *P. bahamense* offers an even easier alternative - that of monitoring the ratio between living and empty cysts through time. Since the *P. bahamense* cyst wall will not degrade after germination (it is fossilizable) and is easily recognizable even when empty, the deposition of new cysts and the germination of living cysts would be seen as increases and decreases in the ratio through time. This approach has been used to study the cyst dynamics of *Gonyaulax polyedra* in Scottish sea lochs (Lewis et al. 1985; Lewis 1988), and is already being used in a study of *P. bahamense* in the Samar Sea of the Philippines (K. Matsuoka, pers. comm.).

The need for *P. bahamense* cyst mapping surveys is apparent. No distributional data of any type have been collected, yet the pattern of species dispersal throughout the Indo-West Pacific region may reflect cyst transport as much as the movement of established blooms between countries by surface currents. Similarly, the origin of blooms remains unknown in all affected countries. Do the blooms originate in localized areas nearshore, and then move with currents to adjacent locations? Or alternatively, are there offshore seedbeds (as is the case with *A. tamarense* in some regions (Anderson and Keafer 1985) that can initiate blooms that are then transported towards shore? These important questions can be addressed in part through cyst mapping surveys.

Discussion

Although the cyst of *P. bahamense* has been known for twenty years, suprisingly little is known of its germination requirements, physiological tolerances and geographic distribution. The research issues described above should be investigated in any program attempting to understand *P. bahamense* bloom dynamics, but the results must be interpreted with caution. The fact that *P. bahamense* has a cyst indicates that the cells necessarilly spend some of their existence in a dormant state in the sediment. What is not clear is how important those cysts are in general bloom dynamics. In temperate waters where cysts provide a mechanism for over-wintering, the dormancy strategy is of great importance if a species is to persist from year to year. In tropical waters where temperature fluctuations are less severe, encystment may provide other more subtle benefits such as those arising from genetic recombination. It is conceivable that careful scrutiny of the water would find at least a few *P. bahamense* motile cells present at all times, with the cyst inoculum being relatively insignificant compared to the growth of the low level vegetative population. This speculation is probably incorrect, but is offered to emphasize that cysts can be important for different reasons in different regions.

Acknowledgements

This work was supported in part by the National Science Foundation (OCE-86-149210) and by the Donaldson Charitable Trust. Contribution No. 7128 from the Woods Hole Oceanographic Institution.

References

Anderson, D.M. and F.M.M. Morel. 1979. The seeding of two red tide blooms by the germination of benthic *Gonyaulax tamarensis* hypnocysts. Est. Coast. Mar. Sci. 8: 279-293.

Anderson, D.M., D.M. Kulis, J.A. Orphanos and A.R. Ceurvels. 1982. Distribution of the toxic red tide dinoflagellate *Gonyaulax tamarensis* in the southern New England region. Est. Coast. Shelf Sci. 14: 447-458.

Anderson, D.M., D.G. Aubrey, M.A. Tyler and D.W. Coats. 1982. Vertical and horizontal distributions of dinoflagellate cyst in sediments. Limnol. Oceanogr. 27: 757-765.

Anderson, D.M. 1984. The roles of dormant cysts in toxic dinoflagellate blooms and shellfish toxicity, p. 125-138. *In* E. Ragelis (ed.) Seafood Toxins. Am. Chem. Soc. Symposium Series. Washington, D.C.

Anderson, D.M., D.M. Kulis and B.J. Binder. 1984. Sexuality and cyst formation in the dinoflagellate *Gonyaulax tamarensis:* Cyst yield in batch cultures. J. Phycol. 20: 418-425.

Anderson, D.M. and B.A. Keafer. 1985. Dinoflagellate cyst dynamics in coastal and estuarine waters, p. 219-224. *In* D.M. Anderson, A.W. White and D.G. Baden (eds.) Toxic Dinoflagellates. Elsevier, New York.

Anderson, D.M., D.W. Coats and M.A. Tyler. 1985. Encystment of the dinoflagellate *Gyrodinium uncatenum:* temperature and nutrient effects. J. Phycol. 21: 200-206.

Anderson, D.M. and B.A. Keafer. 1987. An endogenous annual clock in the toxic marine dinoflagellate *Gonyaulax tamarensis*. Nature 325: 616-617.

Anderson, D.M., C.D. Taylor and E.V. Armbrust. 1987. The effects of darkness and anaerobiosis on dinoflagellate cyst germination. Limnol. Oceanogr. 32: 340-351.

Beam, C.A. and M. Himes. 1980. Sexuality and meiosis in dinoflagellates, p. 171-206. *In* M. Levandowsky and S.H. Hutner (eds.) Biochemistry and Physiology of Protozoa. 2nd Ed. Vol. 3, Academic Press, London.

Binder, B.J. and D.M. Anderson. 1987. Physiological and environmental control of germination in *Scrippsiella trochoidea* (Dinophyceae) resting cysts. J. Phycol. 23: 99-107.

Blackburn, S.I., G.M. Hallegraeff and C.J. Bolch. 1989. Vegetative reproduction and sexual life cycle of the toxic dinoflagellate *Gymnodinium catenatum* from Tasmania, Australia. J. Phycol. 25: 577-590.

Lewis, J., P. Tett and J.D. Dodge. 1985. The cyst-theca cycle of *Gonyaulax polyedra (Lingulodinium machaerophorum)* in Creran, A Scottish West Coast sea-loch, p. 85-90. *In* D.M. Anderson, A.W. White and D.G. Baden (eds.) Toxic Dinoflagellates. Elsevier, New York.

Lewis, J. 1988. Cysts and sediments: *Gonyaulax polyedra (Lingulodinium machaerophorum)* in Loch Creran. J. Mar. Biol. Ass. U.K. 68: 701-714.

Matsuoka, K., Y. Fukuyo and C.L. Gonzales. 1989. A new discovery of cysts of *Pyrodinium bahamense* var. *compressum* from the Samar Sea, Philippines, p. 301-304. *In* T. Okaichi, D.M. Anderson and T. Nemoto (eds.) Red Tides: Biology, Environmental Science and Toxicology. Elsevier, New York.

Pfiester, L.A. 1975. Sexual reproduction of *Peridinium cinctum* f. *ovoplanum* (Dinophyceae). J. Phycol. 11: 259-265.

Pfiester, L.A. and D.M. Anderson. 1987. Dinoflagellate life cycles and their environmental control, p. 611-648. *In* F.J.R. Taylor (ed.) The Biology of Dinoflagellates. Blackwell Scientific Publications, Ltd., Oxford.

Rossignol, M. 1961. Analyse pollinique de sediments marins Quaternaires en Israel II. Sediments Pleistocenes. Pollen Spores 4: 121-148.

Tyler, M.A., D.W. Coats and D.M. Anderson. 1982. Encystment in a dynamic environment: deposition of dinoflagellate cysts by a frontal convergence. Mar. Ecol. Prog. Ser. 7: 163-178.

Turpin, D.H., P.E.R. Dobel and F.J.R. Taylor. 1978. Sexuality and cyst formation in Pacific strains of the toxic dinoflagellate *Gonyaulax tamarensis*. J. Phycol. 14: 235-238.

von Stosch, H.A. 1973. Observation on vegetative reproduction and sexual life cycles of two freshwater dinoflagellates, *Gymnodinium pseudopalustre* and *Woloszynskia apiculata* sp. nov. Br. Phycol. J. 8: 105-134.

Wall, D. and B. Dale. 1969. The "hystrichosphaerid" resting spore of the dinoflagellate *Pyrodinium bahamense* Plate, 1906. J. Phycol. 5: 140-149.

Wall, D. 1971. Biological problems concerning fossilizable dinoflagellates. Geoscience and Man. III: 1-15.

White, A.W. and C.M. Lewis. 1982. Resting cysts of the toxic, red tide dinoflagellate *Gonyaulax* excavata in Bay of Fundy sediments. Can. J. Fish. Aquat. Sci. 39: 1185-1194.

Watanabe, M.M., M. Watanabe and Y. Fukuyo. 1982. Encystment and excystment of red tide flagellates. I. Induction of encystment of *Scrippsiella trochoidea*, p. 27-42. *In* Nat. (Japan) Inst. Environ. Stud., Res. Rep. No. 30. Eutrophical and Red Tides in the Coastal Marine Environment.

Yentsch, C.M., C.M. Lewis and C.S. Yentsch. 1980. Biological resting in the dinoflagellate *Gonyaulax excavata*. Bioscience 30: 251-254.

Yoshimatsu, S. 1981. Sexual reproduction of *Protogonyaulax catenella* in culture: I. Heterothallism. Bull. Plank. Soc. Japan 28 131-139.

Occurrence of the Cyst of *Pyrodinium bahamense* var. *compressum* in Surface Sediments of Brunei Bay

KAZUMI MATSUOKA
Department of Geology
Faculty of Liberal Arts
Nagasaki University
1-14 Bunkyo-machi
Nagasaki 852, Japan

YASUWO FUKUYO
Department of Fisheries
Faculty of Agriculture
University of Tokyo
1-1-1 Yayoi, Bunkyo-ku
Tokyo 113, Japan

MATDANAN H. JAAFAR
M.W.R.N. DE SILVA
Department of Fisheries, Brunei Darussalam
P.O. Box 2161
Bandar Seri Begawan
Brunei Darussalam

MATSUOKA, K., Y. FUKUYO, M.H. JAAFAR and M.W.R.N. De SILVA. 1989. Occurrence of the cyst of *Pyrodinium bahamense* var. *compressum* in surface sediments of Brunei Bay, p. 89-95. *In* G.M. Hallegraeff and J.L. Maclean (eds.) Biology, epidemiology and management of *Pyrodinium* red tides. ICLARM Conference Proceedings 21, 286 p. Fisheries Department, Ministry of Development, Brunei Darussalam, and International Center for Living Aquatic Resources Management, Manila, Philippines.

Abstract

Cysts of a few dinoflagellate species, including those of *Pyrodinium bahamense* var. *compressum,* which has caused several toxic red tides in Brunei Bay, were discovered in surface sediments of this region. Since their density was very low, a palynological processing technique of chemical extraction was used to isolate the cysts. This is a new discovery of cysts of *P. bahamense* var. *compressum* in Brunei Bay and the second record in Southeast Asia in addition to the Samar Sea, Philippines.

Introduction

The cysts of *Pyrodinium bahamense* var. *compressum* were first found in surface sediments of the Samar Sea in the Philippines. Their significance for the toxic red tides caused by this species was briefly discussed in Matsuoka et al. (1989a).

A training workshop concerning the cyst morphology of *Pyrodinium bahamense* var. *compressum* was held under our supervision on 26 and 27 May 1989 at the Fish Landing Complex of the Department of Fisheries in Muara, Brunei Darussalam, during the Management and Training Workshop on *Pyrodinium* Red Tides. In the workshop, we collected surface sediments at several stations near Muara and Sipitang in Brunei Bay by a TFO gravity corer for detailed observation of the cysts, because the toxic red tides caused by this species have been recorded several times around the west coast of Sabah (Roy 1977; Maclean 1979) and Brunei (Beales 1976). Thus, we expected an occurrence of the cysts of *P. bahamense* var. *compressum.* We were unable to find any cyst in the material during the training workshop using the processing method of Matsuoka et al. (1989b) without sonification. Later the senior author (K.M.) tried to find the cysts using an alternative processing method that is usually adopted by palynologists. The results and an ecological note on the cysts are given in this paper.

Sampling Location and Materials

Off Muara and Sipitang in Brunei Bay (Fig. 1), surface sediments were collected by a TFO gravity corer. The sediments were yellowish grey in color, and sandy silt, silty sand or silt in grain size, and without reducing layers at the surface.

Processing Method

The material, sieved with 125 µm and 37 µm-screens, contained a large amount of calcareous and siliceous biogenic and inorganic

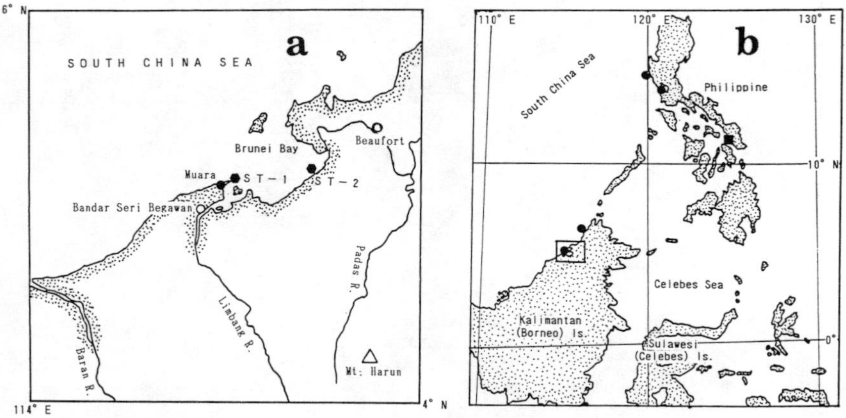

Fig. 1. Sampling locations in Brunei Bay (a), and previous records of *Pyrodinium* bahamense var. compressum (b, ● : motile cell, ■ : cyst).

grains. It was very difficult to find such organic-walled grains as pollen grains, spores and dinoflagellate cysts.

Palynological Technique

At the laboratory in Japan, the upper 2 cm of the original sediment core were placed in a 100 ml silicone beaker; 20 ml of 5% HCl was added for removing such calcium carbonate particles as foraminifers, fragments of shells and corals, and others. After washing with pure water a few times, 10 ml of KOH solution was added to take out fumic acid derived from plant tissues. After rinsing with pure water again, 30% hydrofluoric acid was added to remove fine silicate particles such as diatom valves and other silicate mineral grains. The residual materials were sieved through 120 μm and 20 μm-screens and mounted on a glass slide with glycerine jelly. The observations were carried out under a microscope with interference differential contrast optics.

Results and Discussion

The palynomorphs comprising organic-walled grains such as pollen, spores of ferns, musci and fungi, dinoflagellate cysts and other resting spores of algae, occurred together abundantly with a large amount of plant tissues. Among them, fern spores were most

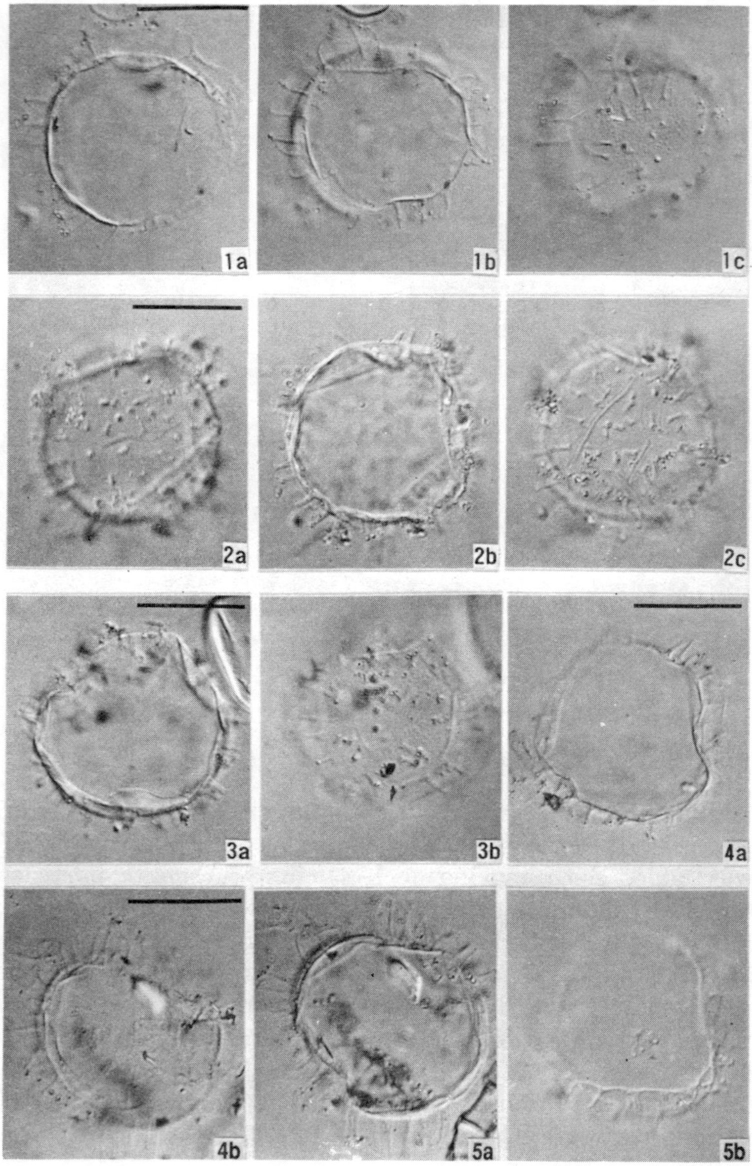

Plate 1

1-5. Cysts of *Pyrodinium bahamense* var. *compressum (Polysphaeridium zoharyi)*, 1: polar view, showing granulate surface (c); loc. Off Sipitang, 2: polar view, showing archeopyle sutures (c); loc. Off Sipitang, 3: polar view, showing epicystal archeopyle (a); loc. Off Muara, 4: polar view, showing epicystal archeopyle (b); loc. Off Sipitang, 5: oblique lateral view, showing archeopyle sutures; loc. Off Sipitang.

Scale bar: 20 µm

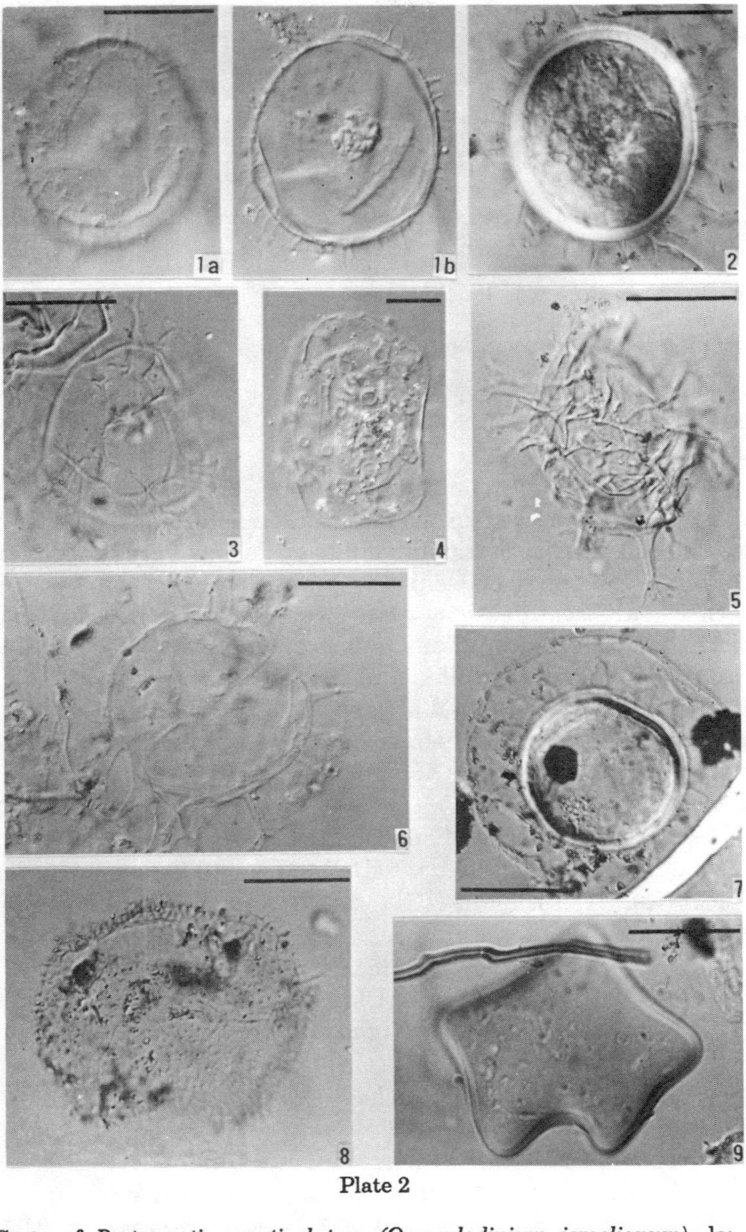

Plate 2

1-2. Cysts of *Protoceratium reticulatum (Operculodinium israelianum),* la: dorsal surface, showing precingular archeopyle, 1b: optical cross section; loc. Off Muara, 2: living cyst filled with protoplasm; loc. Off Muara.

3. Cyst of *Gonyaulax scrippsae* (?) *(Spiniferites bulloideus),* dorsal surface; loc. Off Muara.

(Caption continued next page)

Plate 2. Continued

4, 7. Cysts of *Pyrophacus steinii (Tuberculodinium vancampoae)*, 4: lateral view; loc. Off Muara, 7: optical cross section in polar view; living cyst; loc. Off Muara.

5. Cyst of *Gonyaulax* sp. (*Spiniferites* sp.), loc. Off Muara.

6. Cyst of *Gonyaulax spinifera* complex *(Spiniferites hyperacanthus)*, loc. Off Muara.

8. Cyst of *Protoperidinium pentagonum (Trinovantedinium capitatum)*, dorsal surface, showing intercalary archeopyle; loc. Off Muara.

9. Cyst of *Protoperidinium oblongum? (Votadinium carvum?)*, dorso-ventral view; loc. Off Muara.

Scale bar for 1-3, 5-9: 20 µm, scale bar for 4: 10 µm.

abundant followed by pollen grains, dinoflagellate cysts and fungal spores. The cell density of dinoflagellate cysts was very low, less than 50 cells per ml of wet sediment. The density of *Pyrodinium bahamense* var. *compressum* cysts was 18 cells per ml of wet sediment off Sipitang. The relative number of cysts was approximately 2% of total pollen and spores.

The following dinoflagellate species were identified in the present samples. Biological species names are given first, followed by paleontological synonyms (*).

Gonyaulacacean cysts:

Gonyaulax scrippsae Kofoid

**Spiniferites bulloideus* (Deflandre and Cookson) Sarjeant

Gonyaulax spinifera (Claparede and Lachmann) Diesing complex

**Spiniferites hyperacanthus* (Rossignol) Sarjeant

Protoceratium reticulatum Claparède and Lachmann

**Operculodinium israelianum* (Rossignol) Wall

Lingulodinium polyedra (Stein) Dodge

**Lingulodinium machaerophorum* (Deflandre and Cookson) Wall

Pyrophacus steinii (Schiller) Wall and Dale

**Tuberculodinium vancampoae* (Rossignol) Wall

Pyrodinium bahamense Plate var. *compressum* (Böhm) Steidinger, Tester and Taylor

**Polysphaeridium zoharyi* (Rossignol) Bujak et al.

Peridiniacean cyst:

Protoperidinium spp.

**Brigantedinium* spp.

Protoperidinium oblongum (Aurivillius) Balech?
**Votadinium carvum* Reid?
Protoperidinium pentagonum (Gran) Balech?
**Trinovantedinium capitatum* Reid?
Diplopelta parva (Abe) Matsuoka

In comparison with the subtropical Ishigaki Islands of Japan and the Samar Sea (unpublished data) in the West Pacific, the dinoflagellate cyst assemblage of Brunei Bay is characterized by a low cyst density and a very low species diversity, in particular of peridiniacean cysts. This is probably due to high sedimentation resulting from discharge of rivers, and a low diversity of thecate species producing resting cysts. The cysts of *Pyrodinium bahamense* var. *compressum* occurred both off Muara and Sipitang in Brunei Bay. Most of the cysts were empty with an epicystal archeopyle. The presence of the empty cysts suggests that these cysts probably succeeded in germination and played a role as a seed population for a forthcoming outbreak of *Pyrodinium bahamense* var. *compressum.*

Acknowledgements

We are grateful to Dr. W. Booth and Mr. G. Usup for their kind help to collect samples and to collect samples and to Drs. G.M. Hallegraeff and J.L. Maclean for fruitful discussion concerning *Pyrodinium* cysts and red tides. One of the authors (Y.F.) was financially supported by the Ministry of Education, Science and Culture of Japan in a Grant-In-Aid for Scientific Research Fund No. 01041066, and financial assistance to K.M. was partly by the Ocean Research Institute of the University of Tokyo.

References

Beales, R.W. 1976. A red tide in Brunei waters. Brunei Mus. J. 3(4): 167-182.

Maclean, J.L. 1979. Indo-Pacific red tides, p. 173-178. *In* D.L. Taylor and H.H. Seliger (eds.) Toxic dinoflagellate blooms. Elsevier, North Holland.

Matsuoka, K., Y. Fukuyo and C.L. Gonzales. 1989. A new discovery of cysts of *Pyrodinium bahamense* var. *compressum* from the Samar Sea, Philippines, p. 301-304. *In* T. Okaichi, D.M. Anderson and T. Nemoto (eds.) Red tides: biology, environmental science and toxicology. Elsevier Science Publishing, New York.

Matsuoka, K., Y. Fukuyo and D.M. Anderson. 1989b. Methods for modern dinoflagellate cyst study, p. 461-479. *In* T. Okaichi, D.M. Anderson and T. Nemoto (eds.) Red tides: biology, environmental science and toxicology. Elsevier Science Publication, New York.

Roy, R.N. 1977. Red tide and outbreak of paralytic shellfish poisoning in Sabah. Med. J. Malay. 31(3): 247-251.

Pyrodinium bahamense var. *compressum* Red Tide Studies in Sabah, Malaysia

GIRES USUP
ASMAT AHMAD
NORHADI ISMAIL
Marine Sciences Department
UKM Sabah Campus
Beg Berkunci No. 62, 88996
Kota Kinabalu, Sabah, Malaysia

USUP, G., A. AHMAD and N. ISMAIL. 1989. *Pyrodinium bahamense* var. *compressum* red tide studies in Sabah, Malaysia, p. 97-110. *In* G.M. Hallegraeff and J.L. Maclean (eds.) Biology, epidemiology and management of *Pyrodinium* red tides. ICLARM Conference Proceedings 21, 286 p. Fisheries Department, Ministry of Development, Brunei Darussalam, and International Center for Living Aquatic Resources Management, Manila, Philippines.

Abstract

A study has been carried out since 1985 on the biology and ecology of *Pyrodinium bahamense* var. *compressum* red tides in Sabah, Malaysia. It was observed that the red tide outbreaks tended to occur in June-July and December-January annually, particularly in Brunei Bay, Kimanis Bay and Kota Kinabalu. The onset of these outbreaks coincided with the onset of the northeasterly and southwesterly monsoons. However, in 1988 and 1989, expected outbreaks did not occur, most likely due to heavy rain which seemed to have a negative effect on red tide formation. It is now clear that a period of calm and fine weather of about two weeks is needed for a bloom to develop successfully. Laboratory observations have shown that *P. bahamense* var. *compressum* was aggregatory, underwent vertical migration and emitted fluorescence. The tintinnid *Favella* sp. was found to graze on the dinoflagellate, and the benthic resting cyst of *P. bahamense* var. *compressum* has now also been found in the coastal sediment of Sabah.

Introduction

The Malaysian state of Sabah was one of the first regions in Southeast Asia to face the problem of *Pyrodinium bahamense* var.

compressum red tides associated paralytic shellfish poisonings (Roy 1977; Maclean 1979). It has now been more than a decade since the first outbreak in 1976 and more than 30 fatal cases of PSP have occurred. Yet our knowledge of the organism and its blooming activities remains sketchy and scattered. For example, it was only recently that the benthic cyst of the organism was discovered (Matsuoka et al. 1989), despite the fact that the cyst of the related *P. bahamense* var. *bahamense* was discovered a long time ago (Wall and Dale 1969). With the surfacing of *P. bahamense* var. *compressum* red tides in other parts of Southeast Asia, more researchers within the region are expected to work on the problem. In this paper we summarize the results obtained from our studies on *P. bahamense* var. *compressum* red tides started in 1985 (Usup et al. 1987, 1988).

Materials and Methods

The studies have been carried out on the west coast of Sabah (Fig. 1), particularly in Brunei Bay, Kimanis Bay and Kota Kinabalu. During the red tide outbreak in July 1987, a cruise covered the areas from Kimanis Bay to Usukan Bay to the northeast. In January 1988 a similar cruise was carried out in Brunei Bay during the outbreak there. Permanent sampling stations were also set up in Kota Kinabalu and the seawater lagoon in Kinamis Bay (Kuala Penyu).

During the two cruises, plankton samples were obtained by vertical hauls from 3 m to the surface with a 45-µm plankton net fitted with a flowmeter (Hydro-bios, W. Germany). For studies on the vertical distribution of plankton in Brunei Bay, water samples were obtained from various depths using water bottles. In the shallow seawater lagoon in Kimanis Bay, composite water samples from the surface to 2 m depth were obtained by pumping with a submersible electric pump. Subsamples were filtered through 10-µm Nitex mesh. All plankton samples, except those for live studies, were preserved in either Lugol's iodine solution or 4% formalin.

Environmental parameters (temperature, salinity, pH, dissolved oxygen) were measured *in situ* using the Hydrolab Surveyor multiprobe unit (Environmental Data Systems, USA). Cell counts were made using a Sedgewick-Rafter cell on a Nikon inverted microscope. Fluorescence measurement and photomicrography were carried out on a Nikon microscope fitted with epifluorescence and photomicrography systems.

For culture studies, live cells were isolated and passed through two or three washings in membrane-filtered seawater before being transferred into culture media. Cultures were carried out in petri

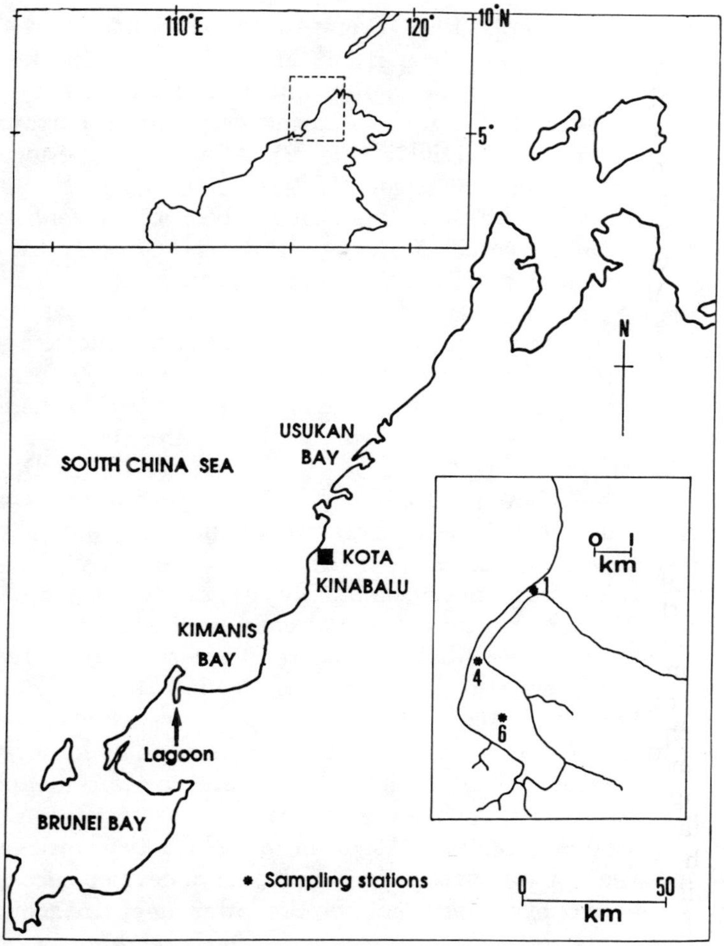

Fig. 1. The west coast of Sabah. Inset: the seawater lagoon in Kimanis Bay showing the sampling stations.

dishes and conical flasks and kept in a controlled environment growth chamber (Forma Scientific, USA) at a temperature of 28°C, and a 12:12 light:dark cycle.

Results and Discussions

Field Observations

Since 1985, we have made observations on five separate *P. bahamense* var. *compressum* red tide events, mainly in Kota

Kinabalu and Kimanis Bay. From these observations we noted certain aspects that were consistent in all blooms. The outbreaks occurred twice a year, in June-July and December-January. These periods coincided with the shift in surface currents to northeasterly and southwesterly, respectively (Fig. 2). Alternatively, the periods can be considered as the onsets of the northeast and southwest monsoons (O'Neill and Eason 1982). These observations lend support to the idea that a period of strong winds is necessary for bloom initiation. The matching of red tide outbreaks from 1976 to 1988 with mean current patterns also indicated that only one outbreak occurred during the transition period between the monsoons. Thus, the general wind and surface current patterns may be important factors in the initiation, spreading and maintenance of red tide blooms in the coastal waters of west Sabah. On this basis, the first red tide in Brunei Bay in 1976 (Maclean 1979) may have started in December 1975 rather than January 1976. Since the outbreaks in Papar and Kota Kinabalu followed the outbreak in Brunei Bay, the cells could have moved out of the Bay with the northeasterly current in December 1975, but the blooms only manifested themselves in January 1976 when the current was already southwesterly.

Whilst windy conditions appeared necessary for red tide outbreaks, rain seemed to inhibit them. In June 1988, and January and June 1989, motile cells of *P. bahamense* var. *compressum* were detected in the water column in Kota Kinabalu (a few hundred cells per liter). However, outbreaks did not occur and the cells disappeared from the water column apparently as a result of heavy rain. Although the exact period is uncertain, it seemed that about two weeks of calm and sunny weather was necessary for a bloom to develop successfully. The adverse effect of rainfall could be the lowering of surface salinity which led to unfavorable growth conditions for *P. bahamense*.

All the blooms that we have observed so far (except in July 1987) have formed visible patches on the sea surface. Even in a limited geographical area, the dinoflagellate can form distinct patches separated by a few hundred meters. For example, in December 1985 five patches were observed in Kota Kinabalu. The patches remained intact for 10 days and drifted with the surface current. We have noted the aggregatory behavior of *P. bahamense* in the laboratory and this could be an important factor in the maintenance of the patches. Additionally, the patches may be formed by convergence due to tidal and other fronts.

Our most extensive monitoring of cell densities during a *Pyrodinium* red tide were in July 1987 (Kimanis Bay to Usukan Bay) and in January 1988 (Brunei Bay). The cell densities in the surface layer (0-3 m) at the sampling stations were as shown in Fig. 3. The

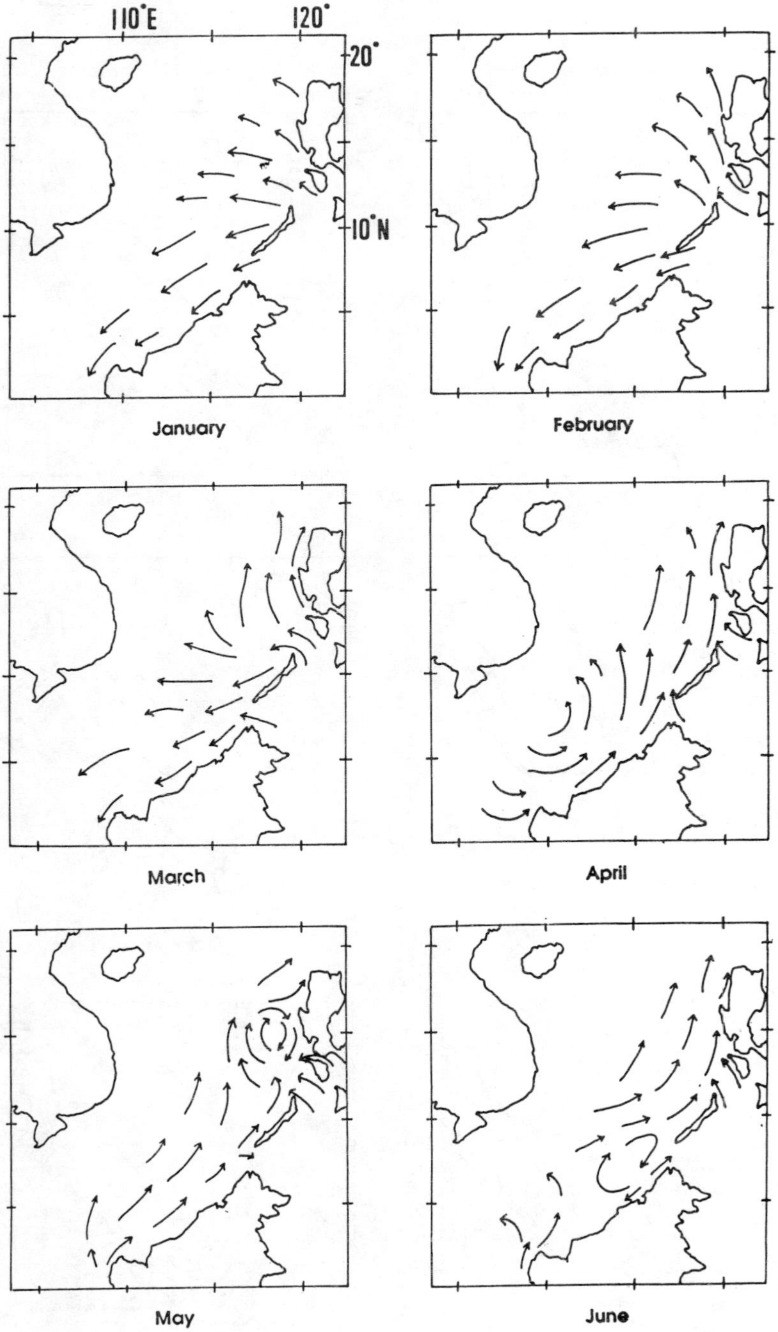

Continued

Fig. 2. General monthly surface current directions on the west and north coasts of Borneo. From: O'Neill and Eason 1982.

Fig. 2. Continued

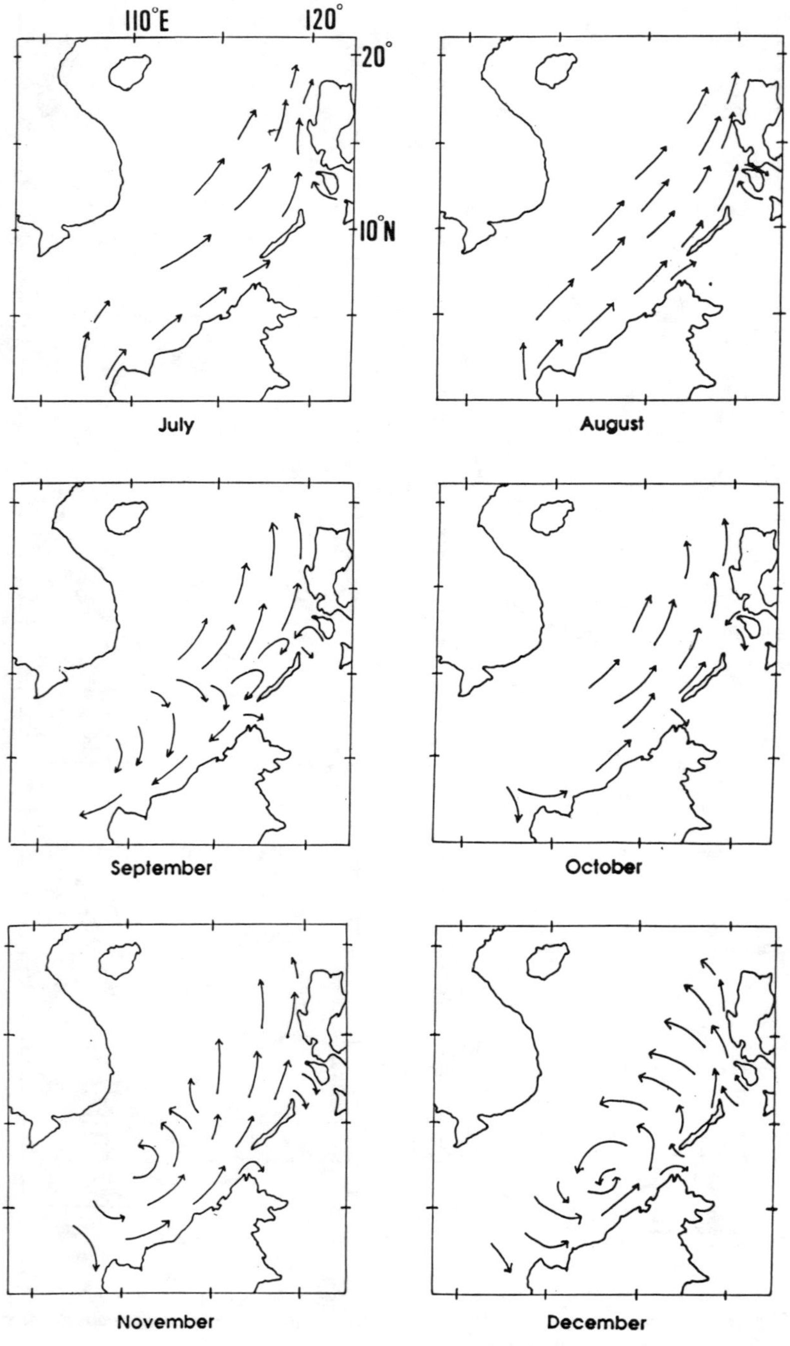

use of a plankton net and flowmeter during the survey introduced a certain amount of error in the quantitative analysis of the data. However, the figures do indicate the magnitude of cell densities which can be present during a red tide. As previously mentioned, the outbreak in 1987 did not form visible patches on the sea surface perhaps due to the choppy sea state during the period. Fig. 3 also shows that cell densities were generally much higher in Kimanis Bay and Kota Kinabalu than at the other stations. The data available so far suggest that these bays are two of the three sites of origin for red tides on the west coast of Sabah, the other being Brunei Bay. Based on this assumption, one can then postulate that the cells encountered between Kimanis Bay and Kota Kinabalu and to the northeast of Kota Kinabalu were transported by the northeasterly currents. The outbreak in Brunei Bay in January 1988 was confined to that area only, and this was most likely due to the predominant southwesterly current during that period. However, this resulted in the spreading of the red tide to the Brunei Darussalam side of the Bay.

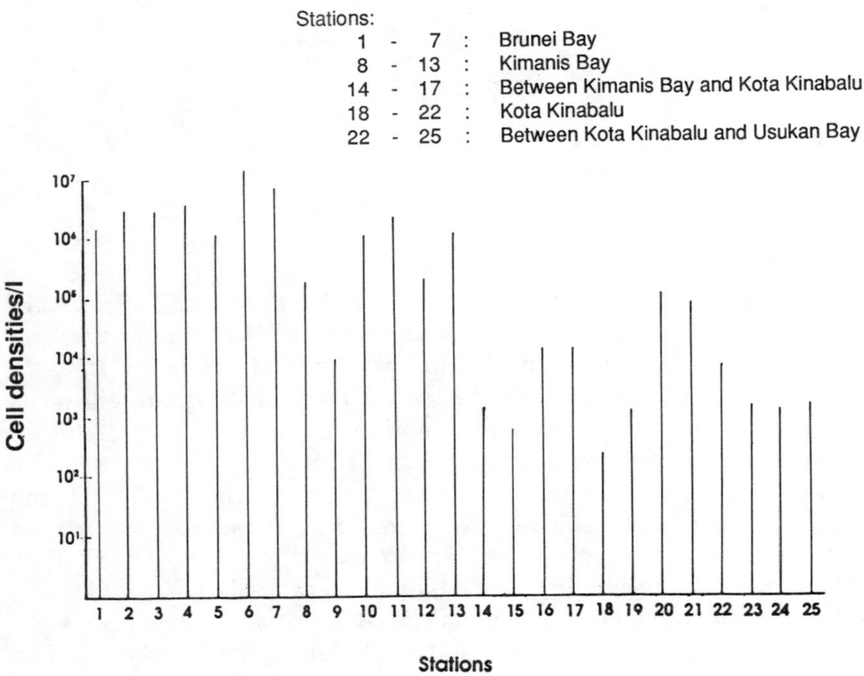

Fig. 3. Cell densities of *P. bahamense* var. *compressum* at 25 stations during the red tides in July 1987 (stations 8-25) and January 1988 (stations 1-7).

During the outbreak in Brunei Bay the depth distributions of dinoflagellate cell densities were as shown in Fig. 4. Cell density was highest in the 3-6 m layer. The lower cell density at the surface was most probably due to the presence of lower salinity water on the surface. The lower cell densities at depths were mainly due to light limitations, probably an effect of self-shading by the high cell densities in the upper layers.

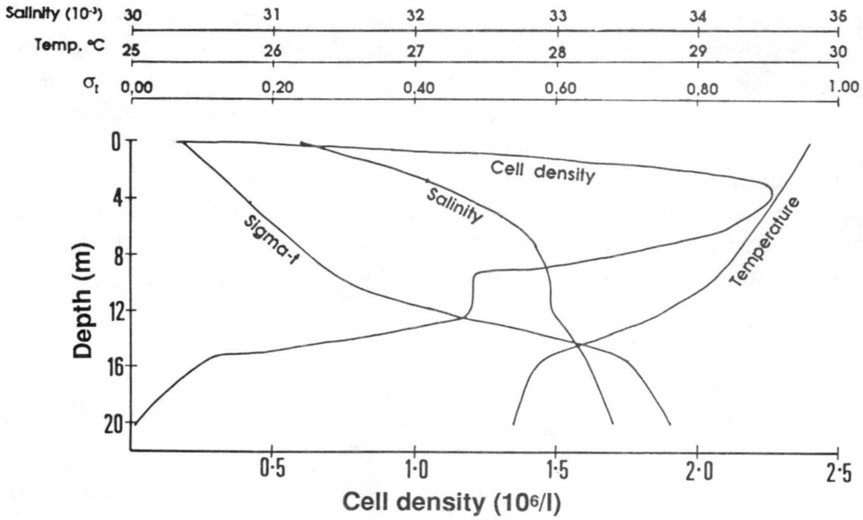

Fig. 4. Depth distribution of *P. bahamense* var. *compressum* cell densities during the January 1988 red tide in Brunei Bay.

In most of the red tides in Sabah, the *P. bahamense* blooms were preceded by the presence of high numbers of *Trichodesmium* and salps in the water column. During this period motile cells of *P. bahamense* could already be detected in the water column, either as single cells or in short chains. Within five or six days, the *P. bahamense* red tide would have approached its peak, which normally lasted 3 to 4 days. During the peak period, about 95% of the phytoplankton in the bloom patch was *P. bahamense,* the others being *Ceratium furca* and *C. lineatum.* During this period *P. bahamense* cells were always in chains several cells long, indicating high division rates. Of the zooplankton, only calanoid copepods were present, usually in quite high numbers. These copepods may graze on *P. bahamense,* as has been reported for other dinoflagellates (Sellner and Olson 1985). *P. bahamense* var. *compressum* cells were further noted to undergo vertical migration, being concentrated on the surface from about 1000 to 1500 hours. They emitted a bluish-white fluorescence when agitated.

After about four days, the red tide would start to decline, and species diversity would become higher. During this period the tintinnid *Favella* sp., also became more abundant in the water column. This tintinnid can graze actively on *P. bahamense* both in the field and in the laboratory (Fig. 5). The decline of *P. bahamense* var. *compressum* blooms was rapid, most likely due to the combined effects of factors including nutrient exhaustion, increased grazing and the destruction of water column structure. A *P. bahamense* bloom was often followed by a bloom of the diatom *Chaetoceros* sp.

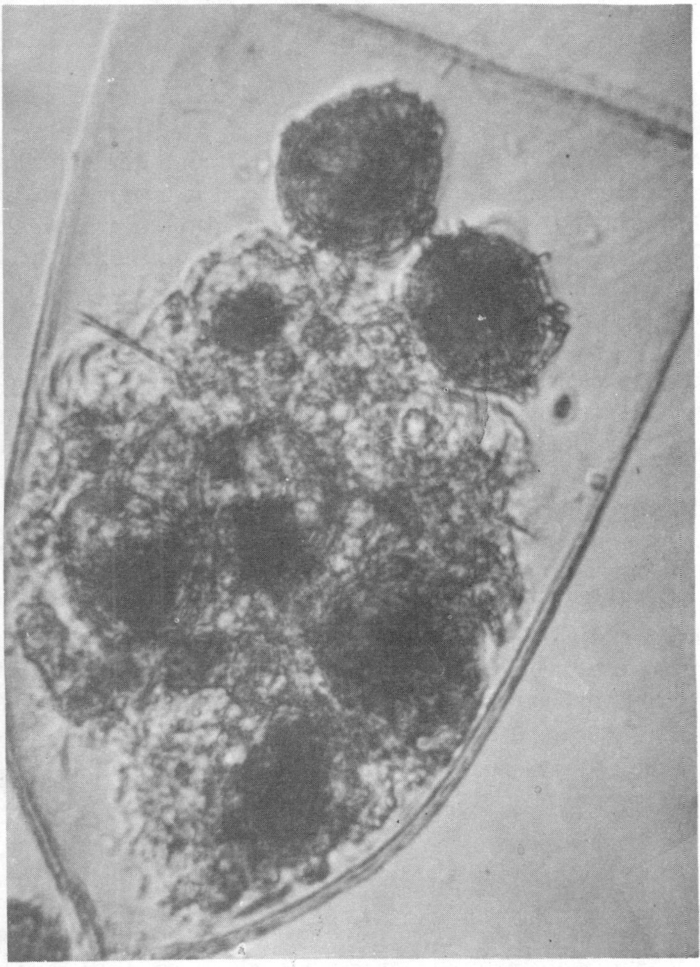

Fig. 5. The tinnid *Favella* sp. containing partially digested cells of *P. bahamense* var. *compressum* (magnification 400 x).

In the coastal waters of west Sabah, the disappearance of a bloom was followed by a period when no motile *P. bahamense* cells were detected in the water column. However, in a protected, semi-enclosed body of water like the lagoon in Kimanis Bay, our study from August 1986 to January 1987 showed that motile cells of *P. bahamense* were present in the water column throughout the study period (Fig. 6). The cell densities were generally low, never reaching the peak values attained during blooms in open coastal waters. The cell densities peaked in August and November. However, the exact peak period could change from year to year depending upon the prevailing weather conditions. The existing situation makes the lagoon and adjacent waters unsuitable for bivalve culture.

Laboratory Observations

So far we have not succeeded in culturing the dinoflagellate. Several published media were tested either in original or modified form, but whilst the cells stayed alive for up to several weeks, they never divided. Nevertheless, some information on the behavior of the cells was obtained. Even in homogeneous media in culture vessels, the cells tended to aggregate, forming a patch. Vertical migrations of the cells in the culture vessels were also noted, although the rhythm seemed to be lost after about five days.

Within a day or two after isolation into a growth medium, most of the cells underwent ecdysis. The cells then assumed a rounded form with a distinct cingulum (Fig. 7). This ecdysal stage was never found in any of the natural samples collected. In culture, these cells were not able to form new thecae, which could be an indication that the medium was unsuitable for growth.

A final observation we made in our studies was that *P. bahamense* can survive for quite a lengthy period in the dark. Water samples from a bloom patch were collected into black, 20-liter plastic carboys and kept at room temperature in the laboratory. After a period of between 10 to 14 days, it was found that the cells in the water samples were not only alive, but were highly motile, thecate and some were even in short chains.

Cyst/Benthic Studies

Recently, we started to study benthic cysts of *P. bahamense,* particularly their distribution and germination. The first area sampled was the lagoon in Kimanis Bay, and intact cysts of the

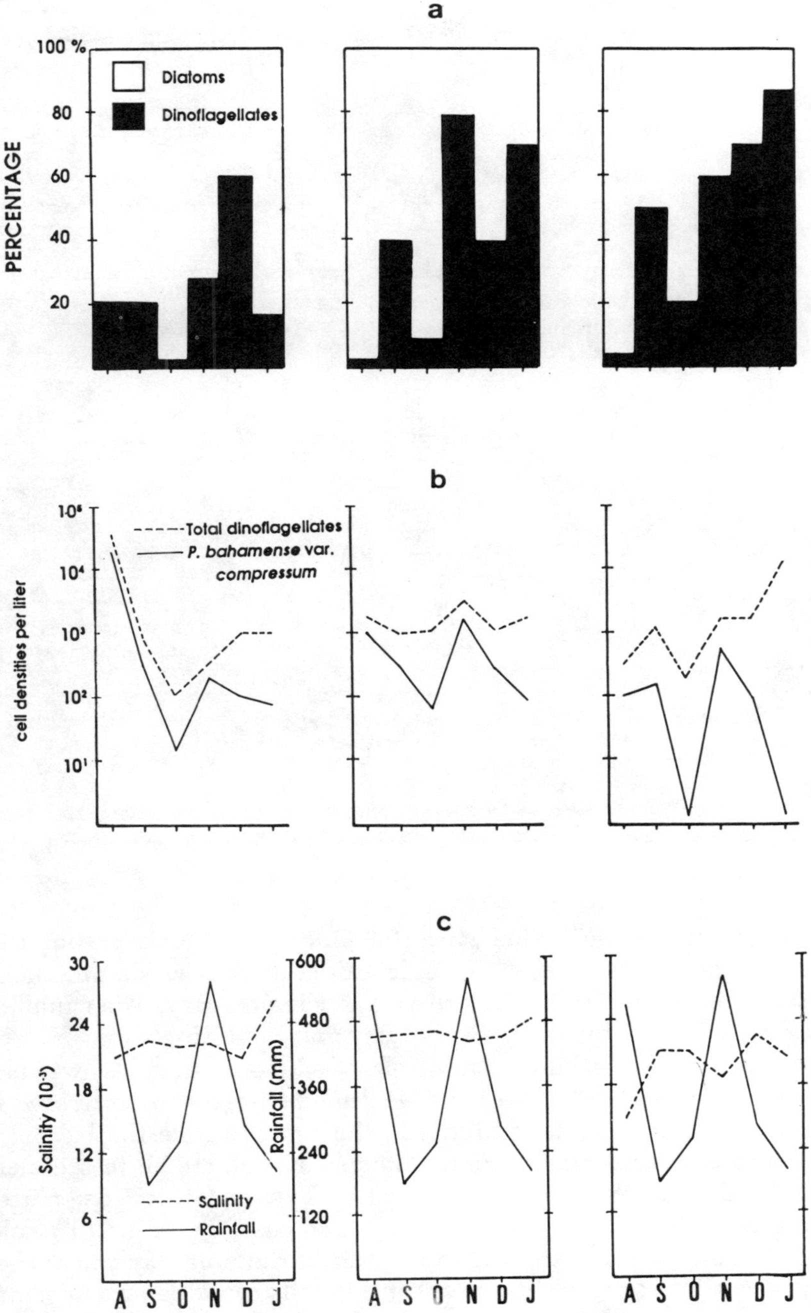

Fig. 6. Relative abundance of diatoms and dinoflagellates (a), *P. bahamense* var. *compressum* density (b), and salinity and rainfall (c) at 3 stations in the seawater lagoon in Kimanis Bay from August 1986 to January 1987.

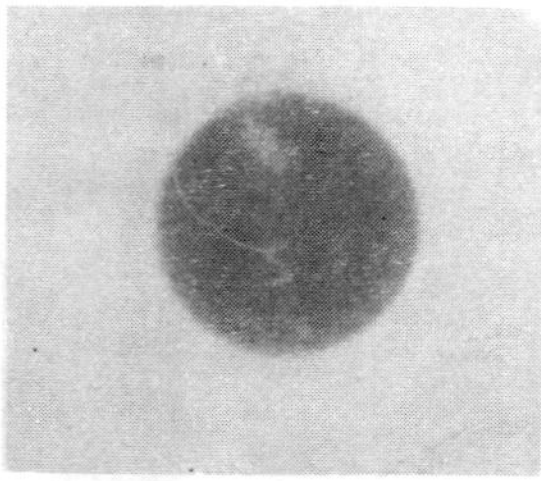

Fig. 7. A round, motile cell resulting from the ecdysis of a *P. bahamense* var. *compressum* cell in culture medium (magnification 200 x).

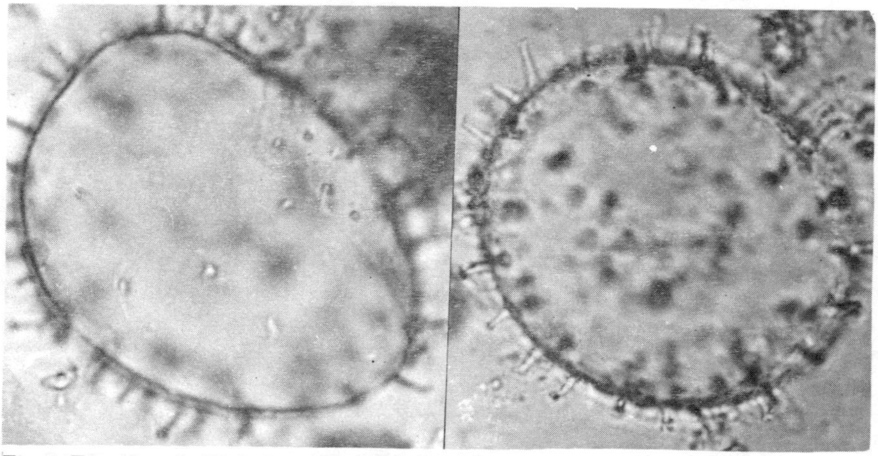

Fig. 8. Empty cyst of *P. bahamense* var. *compressum* from Kimanis Bay sediment.

dinoflagellate in the sediment were found (Fig. 8). Our preliminary data indicate that the cysts were deposited more towards the mouth of the lagoon and on the western side of Kimanis Bay. The sampling program will be extended to Brunei Bay and Kota Kinabalu.

During the examination of the sediment samples we also observed intact vegetative cells of *P. bahamense* var. *compressum* in the surface sediment, particularly in the samples from the lagoon. A similar observation was made by Matsuoka et al. (1989) in sediment samples from the Samar Sea, Philippines. These benthic cells emitted fluorescence, similar to live motile cells and possibly were still viable. In a shallow environment like the lagoon in Kimanis Bay where they can be easily resuspended, these benthic cells could be an important factor in bloom initiation.

The studies which we have conducted so far have by no means brought us any nearer to solving the problems of toxic red tides in Sabah. However, they have provided some basis for predicting when an outbreak is most likely to occur, given the prevailing weather conditions. We have also managed to locate areas where permanent sampling stations for cells of *P. bahamense* var. *compressum* should be established. Further it is anticipated that through our cyst mapping studies we would be able to map the extent of the red tide problem in the coastal waters of Sabah which could be crucial to the shellfish industry in the state.

In summary, we suggest an idealized development curve for a typical *P. bahamense* var. *compressum* red tide in the coastal waters of Sabah (Fig. 9). In the initiation stage, the source of motile cells in a particular water body could either be from benthic cysts or from offshore locations. Environmental factors then determine whether a bloom will be aborted or proceed successfully. The dashed line at the end of the declining phase indicates our uncertainty about events after a bloom. These include the percentage of live cells that formed benthic cysts, the percentage of live cells that died, and the percentage that were transported or dispersed to other locations, such as offshore. We hope that the curve will serve as a useful basis for a future model on red tides.

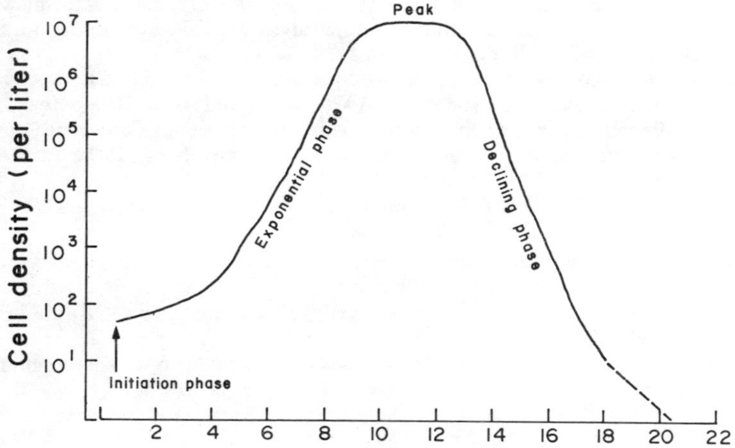

Fig. 9. An idealized growth curve of a typical *P. bahamense* var. *compressum* red tide on the west coast of Sabah.

Acknowledgements

This research project was supported by a Universiti Kebangsaan Malaysia research grant No. 124/85. We would also like to extent our gratitude to the Sabah Fisheries Department and the Sabah Marine Department for the use of their vessels during the research cruises. Similarly we would like to thank the crews of RV Sahabat and MV

Dinawan II for their assistance during the cruises. Finally, we thank various individuals who provided invaluable comments and suggestions during the workshop in Negara Brunei Darussalam. Dr. G. Hallegraeff assisted with the cyst survey in Kimanis Bay.

References

Anderson, D.M. 1984. Shellfish toxicity and dormant cysts in toxic dinoflagellate blooms, p. 125-138. *In* E.P. Ragelis (ed.) Seafood toxins. ACS Symposium Series, American Chemical Society, Washington, DC.

Maclean, J.L. 1979. Indo-Pacific Red tides, p. 173-178. *In* D.L. Taylor and H.H. Seliger (eds) Toxic dinoflagellate blooms. Proceedings of the Second International Conference on Toxic Dinoflagellate Blooms, Florida. Elsevier/North Holland. New York, Amsterdam, Oxford.

Matsuoka, K., Y. Fukuyo and C.L. Gonzales. 1989. A new discovery of cysts of *Pyrodinium bahamense* var. *compressum* from the Samar Sea, Philippines, p. 301-304. *In* T. Okaichi, D.M. Anderson and T. Nemoto (eds.) Red tides: biology, environmental science and toxicology. Proceedings of the International Symposium on Red Tide Biology, Environmental Science and Toxicology, Takamatsu, Japan. Elsevier Science Publishing, New York.

O'Neill, B.G. and C.W. Eason. 1982. China Sea Pilot. Vol. II. Hydrographer of the Navy, Hydrographic Department, Ministry of Defence, Somerset, England.

Roy, R.N. 1977. Red tide and outbreak of paralytic shellfish poisoning in Sabah. Medical Journal of Malaysia 31(3): 247-251.

Sellner, K.G. and M.M. Olson. 1985. Copepod grazing in red tides of Chesapeake Bay, p. 245-250. *In* D.M. Anderson, A.W. White and D.G. Baden (eds.) Toxic dinoflagellates. Proceedings of the Third International Conference on Toxic Dinoflagellates, St. Andrews, Canada. Elsevier. New York, Amsterdam, Oxford.

Usup, G., A. Ahmad, N. Ismail and G. Ismail. 1987. The red tide phenomenon in Sabah. Sabah Society Journal III(3): 320-328.

Usup, G., J. Salimon, A. Ahmad and N. Ismail. 1988. Red tides in Sabah waters, p. 33-50. *In* A. Sasekumar, R.D. Cruz and S.L.L. Hong (eds.) Proceedings of the 11th Annual Seminar, Malaysian Society of Marine Sciences, Kuala Lumpur, Malaysia.

Wall, D. and B. Dale. 1969. The hystrichosphaerid resting spore of the dinoflagellate *Pyrodinium bahamense* Plate 1906. J. Phycol. 5: 140-149.

Research Personnel and Training Needs

Chairperson : T. Okaichi
Rapporteur : J.L. Maclean

As a background to what personnel are needed to investigate red tides, the situation in the Seto Inland Sea, Japan was described. There are two sources of research personnel - universities and the prefectural fisheries experimental research stations. The stations each have 2-3 biologists and 2-3 toxicologists involved in red tide work. A large survey of the Seto Inland Sea is beginning, for which this number of scientists is inadequate. Training of fishermen is being carried out to enable them to differentiate bloom species and collect water samples. This makes up for the shortage of scientists.

Regarding *Pyrodinium* red tides, the situation in the various countries was then described.

- In the Philippines, the Bureau of Fisheries and Aquatic Resources (BFAR) is working closely with several other agencies following the spread of red tides to Manila Bay in 1988. However, there is a lack of manpower, especially considering the increase in red tides in 1987 and 1988. Meanwhile, staff have been cut by 50%. BFAR has four two-person teams for red tide surveys, who have trained over 70 researchers in the provinces to detect red tides and monitor the waters. The Marine Science Institute of the University of the Philippines (UPMSI) has six marine biologists and biochemists and is planning a basic research project on red tides. One UPMSI researcher has trained in Sherwood Hall's FDA laboratory in Washington, as has another from the UP College of Medicine. The Philippines Department of Health works closely with other agencies, especially BFAR. Cooperation with other agencies is necessary to cover the vast area involved.

- The Sabah Fisheries Department badly lacks manpower but only carries out monitoring activities. The university is carrying out limited research.
- In Singapore, a single PSP incident has occurred but *Pyrodinium* has not been detected. The Primary Production Department conducts fortnightly surveys to detect PSP and the occurrence of phytoplankton blooms in relation to fish kills in net cages in the Johore Straits. Main problems are the limited manpower and the inability to identify the causative organisms with confidence. Training in identification of dinoflagellates is required. To date, only one technician from the National University of Singapore has undergone short-term training on surveying phytoplankton blooms, under Max Taylor in the University of British Columbia, Canada.
- Indonesia lacks training and manpower in all aspects. Monitoring is carried out in Jakarta Bay only.
- In Thailand, the Department of Fisheries carries out monitoring. Three persons are doing research at Chulalongkorn University in cooperation with Japanese groups, but not on *Pyrodinium*.
- Brunei Darussalam has been carrying out a comprehensive monitoring program to enable essential policy decisions to activate a Red Tide Action Plan. It has acquired an HPLC and other equipment to upgrade monitoring and research capability.
- Guatemala and Papua New Guinea have a complete lack of training in this field.

It was pointed out that there were a number of areas identified in the session dealing with recommendations for future research, and trained personnel would be needed to carry out the work. The nature and extent of training has to be identified. On the other hand, if the focus is on health aspects, then only monitoring is needed. There is a need to distinguish between academic and applied research.

In general, the present training approach is for individuals to spend a few months in one of just a few laboratories in Canada, Japan or the USA. It is "becoming a burden" to some such laboratories. One alternative is for trainers to spend one or more months in the country concerned. There is no formal postgraduate training in this field available anywhere. One solution would be for a university with expertise to offer 3-6 weeks intensive postgraduate training courses. Since the research and training facilities should be within the one institution, such postgraduate courses should take place in a *Pyrodinium*-affected country. Brunei Darussalam expressed interest in developing a unit for this purpose, as did the Marine Science Institute of the University of the Philippines.

Initially, trainers from outside the region would be needed to man postgraduate courses. Eventually, however, the institution would become independent and be the seat of training for the region.

Finally, the view was expressed that there was no sense in every country attempting to do all the research. The present workshop has provided directions on priority information needs to support red-tide activities in each country.

Discussion and Recommendations on Research on the Biology, Ecology and Toxicology of *Pyrodinium*

Chairperson : D.M. Anderson
Rapporteurs : D.M. Anderson, G.M. Hallegraeff and J.L. Maclean

Discussion was opened with comments on the identity of *Pyrodinium* itself. Are there really subspecies? Is morphological compression in the western Pacific populations an artifact of chain formation? What is the significance of such differences in terms of toxicity? One clear need was for culture studies to test for sexual compatibility as a means to resolve these questions.

However, unialgal dinoflagellate cultures may show variations in toxin profiles, plate formulation, chain formation, biochemistry, and, in the case of *Pyrodinium,* luminescence. Luminescence has rarely been observed by researchers in the western Pacific. It was seen once in Sabah during a bloom as well as in an attempted culture.

Identity of the cysts is also in doubt. Some authorities believe that cysts of the two "subspecies" are different.

The question of spreading of the species was raised in this context. Since cysts of *P. bahamense* were widespread in geological times from Israel to Australia, it would appear that distribution has contracted to the present time. On the other hand, dinoflagellates have been proven to be capable of transfer across oceans in the ballast of ships, for example, between Japan and Tasmania, and recent toxicity history within the Indo-West-Pacific suggests an expansion. Cyst studies could be used to indicate the past history of a species in a region, but they are difficult due to mixing of sediment by currents and grazing animals. It may be possible to use other "marker" species/cysts in recent sediments to date sections of cores, or to use a "natural" isotope such as Cesium 137 which marks the first French atomic tests in 1954 in parts of the Pacific.

It was noted that some core sampling was being done in the Philippines by the Bureau of Fisheries and Aquatic Resources, while the Marine Science Institute of the University of the Philippines was planning to study sediments in several locations. Cores were also obtained from Brunei Bay during the present workshop. However, an important point is that the size and duration of blooms may have no relationship with cyst concentrations in the underlying sediment.

Locations of cyst deposition might vary through time, making interpretation of coring results difficult. Past sea level changes were a cause of such variation. There is also variability due to eutrophication. Sediments in areas where waters are relatively unpolluted typically harbor a healthy benthic community capable of burying cysts well below the sediment surface into anoxic layers where germination is not possible. One intriguing possibility is that the quite different benthic community in polluted or eutrophied areas cannot bury the cysts as deep, thus giving an advantage to cyst-forming species in those waters. This may help to explain the apparent expansion or spreading of certain toxic species as man's influence on the coastal zone increases.

With regard to hydrography, the scale of events is important. The well documented series of events, including a major hurricane, in Massachussetts in 1972 which brought red tides into a new area was cited. The red tides subsequently underwent "dampening" and disappeared in certain embayments at the extreme end of the population distribution. Two factors contribute: the annual transport, which varies with climate and is seasonal; and unusual hydrographic events, which are variable and sporadic.

A broad correlation was suggested between increase in red tides and increase in the coastal population of humans. However, the historical use of shellfish in the Philippines, for example, would suggest that coastal communities should have known about PSP if it were a slowly growing rather than a new acute problem. The absence of PSP from *Pyrodinium* in the Caribbean area might simply be a reflection of the low level of shellfish consumption there.

It was pointed out that the chief vector in Philippine PSP, the mussel *Perna viridis,* only became a popular food item in the 1950s. Furthermore, mussels were only introduced into the Samar area (which has been affected in all Philippine red tides and was the site of the first outbreak in 1983) in 1976 and consumed since 1978. They were introduced at the same time to Iloilo and Bacolod where PSP has occurred recently.

It was felt extremely unlikely that PSP cases in Brunei had occurred before the first outbreak in 1976. It was further noted that *Pyrodinium* was absent in extensive plankton surveys in Malaysia in the late 1960s.

In the Guatemala incident, the donacid clam transvector remained available to harvesters later in the year than usual, into the red tide "season". Perhaps the overlapping of the clam and red tide "seasons" was responsible for the PSP, i.e., red tide was previously present but unnoticed.

Toxicity in fish due to *Pyrodinium* remained an unresolved issue. PSP has been linked to the death of fish-consuming whales, and mackerel livers contained saxitoxin at levels up to 600 μg/100 g of tissue. Saxitoxin is known to accumulate in pufferfish livers. In Brunei, various *Sardinella,* pelagic fish, had intestines full of *Pyrodinium* and many of those cells were ecdysing. It is curious that many of the poisoning cases in Brunei were from fish consumption, whereas in the Philippines where small, planktivorous fish are also eaten, there have been virtually no illnesses due to PSP in fish.

The following research topics emerged as important areas of investigation from the above discussions. They were presented and discussed in a plenary session.

Vegetative Cell Biology, Ecology and Distribution

Virtually nothing is known about the dynamics, spatial distributions and general ecology of *P. bahamense* vegetative cells. Clearly, however, cost-effective monitoring and management strategies require a thorough knowledge of these issues. Accordingly, the following research topics are considered to be of high priority.

- *Bloom distribution.* Considerable uncertainty exists as to the hydrographic environments that are most favorable for *P. bahamense* growth and accumulation. Comprehensive field studies are needed to document the vertical and horizontal distribution of cells within the region. The possible existence of offshore blooms should be investigated, as these may be important sources of toxicity for nearshore shellfish following transport events, as well as in species dispersal within the region. Likewise, the apparent spatial association between this species and mangroves can be investigated with areal surveys.
- *Hydrographic studies.* Concurrent with bloom distribution studies, hydrographic data should be collected to define the water column structure and general circulation patterns associated with the presence or absence of *P. bahamense* blooms.
- *Remote Sensing.* Hydrographic and bloom distribution data should be examined with respect to remote sensing images when they are available. Chlorophyll images may be

misleading since they only reflect surface phytoplankton populations, but infrared data of surface water temperature can be quite useful in indicating important water circulation patterns.

- *Associated Organisms.* During blooms of *P. bahamense,* it is important to identify and count co-occurring organisms, especially competing phytoplankton species and zooplankton.

The apparent spatial association between *P. bahamense* and mangroves should be investigated in laboratory culture experiments. Mangrove communities might provide necessary growth factors similar to those in "soil extract", a common supplement to culture media.

- *General Culture Experiments.* The basic physiological and growth characteristics of *P. bahamense* should be determined in laboratory culture experiments. Growth should be studied as a function of temperature, salinity, light and nutrients, for example. Factors affecting toxin production should also be investigated in culture experiments.
- *Culture collections* of a variety of strains of *P. bahamense* from the Indo-West Pacific and tropical areas in the Atlantic should be established, maintained and characterized taxonomically, biochemically and physiologically.

Cyst Biology, Ecology and Distribution

Pyrodinium bahamense has a resting cyst stage in its life history that can be important in species dispersal, bloom initiation, bloom decline and possibly even in toxin transvection. The following research topics are deemed to be of high priority.

- Determine the life history of *P. bahamense.*
- Conduct studies to determine the dormancy (maturation) requirements of *P. bahamense* cysts, as well as germination characteristics as a function of temperature and light. This knowledge is needed to determine the "seeding" or inoculum potential of cysts at different times of the year or under different environmental conditions.
- Conduct cyst mapping studies. In some instances, a qualitative (presence versus absence) distributional study can be very informative in locating areas with toxic potential due to the presence of a "seed" population. In other instances, quantitative cyst mapping could be used to describe large-scale distributions and to locate potential sites of bloom initiation. Such studies will be especially informative if *P. bahamense* is shown to be a nearshore species without

significant offshore distribution. It should be emphasized that the absence of cysts at a station does not mean that PSP is unlikely there since transport of cells to an area from source waters elsewhere is possible. Cyst surveys pinpoint "seed beds" which can then be monitored.

The durability of the *Pyrodinium* cyst wall permits studies of the relative proportion of full versus empty cysts as an indicator of germination and cyst deposition. This approach has been used successfully in studies of another important red tide dinoflagellate, *Gonyaulax polyedra*.

Toxicology

- Based on experience with *Alexandrium* species, which have been shown to exhibit considerable regional variability in toxin composition, it is important to determine the extent to which similar variability occurs among *P. bahamense* populations. HPLC analyses of bloom samples, shellfish and fish, and cultures of *P. bahamense* isolates should be obtained.
- It is equally important to study the potential for, and actual extent of, conversions of toxins during food chain transfer, cooking or other processing, or following human ingestion. Depending on the mixture of different toxins in a sample, decarbamoylation or acid hydrolysis could result in significant enhancement in toxicity.
- Management of toxic shellfish resources requires a sound understanding of the dynamics of toxin accumulation and elimination (depuration) by different species of molluscs. The effect of environmental variables on these processes should be evaluated, especially in light of the high temperatures of tropical waters, which should accelerate toxin elimination.
- Very useful information on dinoflagellate bloom dynamics can be obtained from shellfish monitoring. For example, long-term data sets on shellfish toxicity can indicate temporal trends that might suggest water movement patterns and bloom transport. In some instances, the ability of shellfish to "integrate" blooms provides more useful information than field plankton sampling surveys. Comprehensive shellfish monitoring programs should thus be established and supported as long-term commitments.
- Develop and evaluate alternative methods for detection of *P. bahamense* toxins in different fish or shellfish.

Taxonomy

Studies of other dinoflagellates have shown that regional populations of a species actually consist of a mixture of subpopulations, each with distinct biochemical and genetic characteristics. Given the apparent spreading of *P. bahamense* within the Indo-West-Pacific region, it is important to compile information relevant to the taxonomy and inter-relationships between subpopulations of *P. bahamense.* This should include detailed taxonomic examination, as well as toxicological, biochemical and molecular characterizations when possible.

To learn more about the possible dispersal of *P. bahamense* within the Indo-West Pacific region and across the Pacific to Guatemala, sexual compatibility studies should be conducted using geographically separate *P. bahamense* isolates.

Recommendations for *Pyrodinium* Monitoring and Research Equipment

Chairperson : D.M. Anderson

As several countries and agencies within those countries begin to establish monitoring and research programs directed towards the threat from *P. bahamense,* there is a clear need for standard equipment and procedures. The following list, developed and discussed during plenary session in the Workshop, describes such equipment, with indications of those items considered essential, as well as those that would be useful but are expensive or complex and thus should only be purchased under "ideal" conditions where funding and skilled personnel are not limited.

Field Sampling

Qualitative Plankton Sampling. Perhaps the simplest tool that should be available to all laboratories is a plankton net. This is useful in screening water samples for the presence of a dangerous phytoplankton species (through vertical net tows), and in concentrating natural populations for toxin extraction and analysis. Net tows are not quantitative for phytoplankton and should not be used for enumeration. A useful mesh size for a net would be 20-28 µm. The net mouth need not be the traditional shape; triangular nets, nets on sleds, etc. can be used effectively.

Another instrument that provides useful (but not necessary) information that is not truly quantitative is the fluorometer. In the field, a flow-through fluorometer connected to a hose/pumping system can provide real-time information on the vertical and horizontal distribution of phytoplankton chlorophyll. Although all phytoplankton are combined together in such analyses (i.e., the data are rarely useful as indicators of the distribution of one species such as *P. bahamense*), the results can be quite useful in identifying subsurface strata where phytoplankton biomass is highest or where hydrography has concentrated cells, such as at frontal convergences. When combined with a small CTD (conductivity, temperature, depth) profiler, fluorescence data are especially informative.

Field microscopes provide the possibility of *in situ* identification. One useful model is the Swift FM-31 (US$420), a small instrument with excellent resolution that can be used with either a battery-powered light source or direct sunlight. A less expensive alternative is a hand lens ($10) and flat capillary tube (Vitro Dynamics, Inc., 114 Beach St., Rockaway NJ 07866 USA) which can be used to visualize cells without fixation in the field and the laboratory. This method has considerably less resolution than the field microscope, but can be useful if the species of interest is plentiful or easily distinguished from other forms.

Quantitative Phytoplankton Sampling. Perhaps the most important constraint to sampling for *P. bahamense* in natural waters is the tendency of this species to aggregate in layers rather than to be dispersed throughout the water column.

When maximum cell concentrations are desired, only a fluorometer or visual inspection by diving can indicate discrete depths for sampling. In most cases, the most useful data will be the average cell concentration in the mixed layer. To obtain such information, an integrated sampler similar to that described by F.J.R. Taylor shoud be employed (see Franks and Anderson, this vol.). Some research or monitoring programs should consider alternative profiling systems consisting of an inexpensive submersible pump (e.g., Little Giant or boat bilge pump) connected to garden hose. By raising or lowering the hose at a steady rate over specific depth intervals (e.g., 5 meters) and subsampling from a bucket containing all water collected over each interval, a coarse profile cell concentrations can be obtained and an average concentration calculated for the portion of the water column of interest (e.g., the mixed layer above the pycnocline). Note that pump profiling systems require DC or AC power, whereas the pipe sampler is entirely manual and is more easily deployed from a small boat.

When integrated samplers are not available, Niskin, Van Dorn, or other closing sampling bottles are adequate. However, the user should recognize that bottle casts at discrete depths can provide misleading pictures of vertical cell concentration profiles since they may not reveal important subsurface features.

Few countries have adequate resources to provide remote sensing images to those concerned with *P. bahamense* bloom distributions. However, such data may be available at a regional level or through the USA or other countries with active imaging programs.

In any case, facilities to receive satellite or aircraft remote sensing information would be a significant asset for those attempting to correlate dinoflagellate blooms with water circulation patterns.

Cyst Sampling. Sediment samples can be collected with a variety of devices. Gravity corers are most convenient, but Van Veen or other dredges can be used if care is taken to retain the top, flocculent surface layer of sediment. Core collection by SCUBA diving is very effective when possible.

For nonquantitative sampling, any device that collects the top few centimeters of sediment is adequate. Old plankton nets or hand-operated pumps connected to hoses are examples of inexpensive ways to obtain sediment. Sediment traps can also be used, but their nonquantitative nature should be recognized.

Environmental Parameters. In a well-financed research or monitoring program, a small CTD System (e.g., Sea Bird) is extremely useful, especially when interfaced with a flow-through or *in situ* fluorometer. When a multiple sensor profiler is not available, small meters such as that sold by YSI Inc. can be used to obtain salinity, temperature or dissolved oxygen profiles. Other useful instruments would include a spherical (scalar) photometer system (e.g., Licor, Inc.) for measuring light profiles, or an autoanalyzer system (e.g., Technicon, Inc.) for nutrient analyses. Some of these instruments and their use are described by Franks and Anderson (this vol.).

Laboratory Studies

Toxin Detection

No special equipment other than a hot plate, small centrifuge (optional), stopwatch, blender and balance are needed for mouse bioassays.

At the other extreme, HPLC systems for toxin analysis are very expensive and complex; they should only be purchased when research or monitoring goals justify the significant investment of money and personnel required to establish, operate and maintain such a system. Other necessary or useful pieces of equipment that should be available to a laboratory working on toxin identification and quantitation include: centrifuges, electrophoresis apparatus, fraction collectors, freeze driers, balances, and column chromatography equipment for purification and isolation.

Biology

Identification and Counting

A good compound microscope is needed for proper identification of cells and cysts. For cell enumeration, an inverted microscope is a useful but not necessary alternative. Settling chambers allow cells in a sample to be concentrated as much as 50-fold prior to counting with the inverted microscope. Otherwise, Sedgewick-Rafter or Palmer-Maloney counting slides can be used for counting (1.0 and 0.1 ml volume, respectively). Here again, flat capillary tubes can be calibrated (by volume) and used to count living cells, even when motile, under a dissecting microscope. The much more expensive particle counters (such as the Coulter Counter) are only useful for laboratory culture studies and even then are not always accurate for some dinoflagellates.

Culture (Cells and Cysts)

At least one lighted, temperature controlled incubator (refrigerator size) is needed for maintenance of dinoflagellate cultures. A second incubator is ideally needed for back-up cultures to ensure long-term survival of culture collections. Experiments may well require a third incubator in which temperature can be varied at will. When room temperature is well-controlled, a shelf in a window out of direct sunlight can provide inexpensive culture space. Other facilities needed for culture work include an autoclave, vacuum pump for filtration, refrigerators, and a laminar flow hood or other sterile area for culture transfers.

For cyst studies, an ultrasonic probe is preferred for sediment processing, although an ultrasonic bath can work almost as well. Long-term storage of sediment samples will require temperature control in an unlighted incubation bath or refrigerator. Sieves for sediment processing can either be the expensive metal variety used by geologists or home-made units consisting of cylindrical sections of 5-cm PVC pipe to which circles of plankton netting are glued (epoxied).

Management of *Pyrodinium* Red Tides in Brunei Darussalam

M.W.R.N. DE SILVA
ABDUL HALIDI MOHD SALLEH
SITI AMIN MAHALI
SELVANATHAN SUBRAMANIAM
Department of Fisheries
P.O. Box 2161
Bandar Seri Begawan 1921
Brunei Darussalam

DE SILVA, M.W.R.N., A.H.M. SALLEH, S.A. MAHALI and S. SUBRAMANIAM. 1989. Management of *Pyrodinium* red tides in Brunei Darussalam, p. 125-134. *In* G.M. Hallegraeff and J.L. Maclean (eds.) Biology, epidemiology and management of *Pyrodinium* red tides. ICLARM Conference Proceedings 21, 286 p. Fisheries Department, Ministry of Development, Brunei Darussalam, and International Center for Living Aquatic Resources Management, Manila, Philippines.

Abstract

Brunei Darussalam has experienced red tide occurrences in 1976, 1980 and in 1988. The causative agent of these red tide incidences is *Pyrodinium bahamense* var. *compressum.* The present management methods strive to minimize economic losses without compromising the safety to public health. A tri-directional approach is adopted by Brunei Darussalam in its efforts to mitigate the effects of *Pyrodinium* red tides. The components are a) monitoring, b) research and c) education.

A Red Tide Action Plan involving primarily the Department of Fisheries, Ministry of Industry and Primary Resources; Department of Medical and Health Services, Ministry of Health; Radio and Television Brunei, Ministry of Communications; and other news media is put into operation to minimize the impact of *Pyrodinium* red tides.

Monitoring Red Tide Organisms and Toxins

Water and Plankton Sampling

The Fisheries Department in Brunei Darussalam has been monitoring the levels of *P. bahamense* var. *compressum* in selected sampling sites in the coastal waters likely to be affected by red tides based on previous experiences. At present 10 stations (Fig. 1): five offshore and five within the inner Brunei Bay, are sampled regularly every two weeks for *Pyrodinium* in vertical plankton hauls and in water samples from 3 different depths.

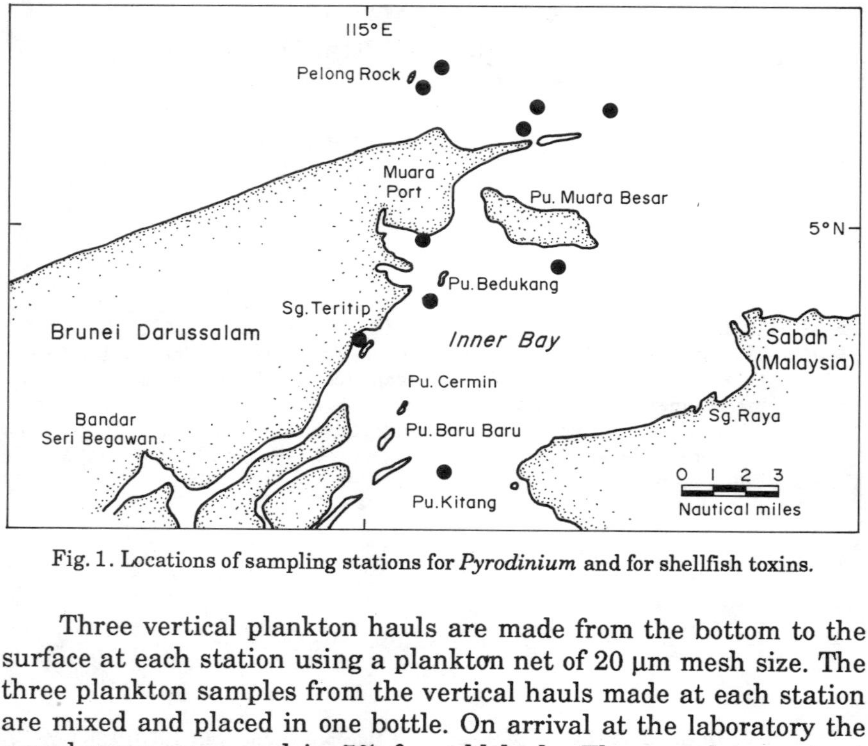

Fig. 1. Locations of sampling stations for *Pyrodinium* and for shellfish toxins.

Three vertical plankton hauls are made from the bottom to the surface at each station using a plankton net of 20 µm mesh size. The three plankton samples from the vertical hauls made at each station are mixed and placed in one bottle. On arrival at the laboratory the samples are preserved in 5% formaldehyde. The samples are kept undisturbed for 3-4 hours for the plankton to settle. The liquid volume is reduced to 10 ml taking care not to disturb the settled plankton. Then 0.01 ml of well shaken preserved plankton sample is scanned under an inverted microscope for the presence of *Pyrodinium* and a count made of all the cells present. Counts are made at least in duplicate and computed to make allowance for the varying depths at each sampling station enabling station to station comparison (shallowest station 4m and the deepest 20 m). Although accurate population counts of *Pyrodinium* might be important from a scientific point of view, even very rough counts of the *Pyrodinium* cells in the plankton haul samples have been found to be sufficient to indicate trends in the population and to initiate further action to mitigate the effects of *Pyrodinium* red tides.

Brunei Darussalam's experience indicates that during what can be considered as *Pyrodinium* red tide "free periods" none or rarely a very few (1 or 2) *Pyrodinium* cells are encountered in the plankton samples. It is of interest to note that from a condition of 0 cells/0.01 ml of plankton haul sample in the sampling stations near Pelong Rocks on 19 January 1988, the counts increased to over 1,500 cells/0.01 ml on 23 January 1988. Whether the sudden increase was due to an influx of *Pyrodinium* from another area or due to the rapid multiplication of cells in the area has not been determined. During the 1988 red tide, a few cells (1 to 10) of *Pyrodinium* were present in the plankton samples during the tail end of the red tide. The presence of *Pyrodinium* cells in the plankton samples is taken as sufficient indication to increase the frequency of *Pyrodinium* cell monitoring from fortnightly to weekly and PSP toxin monitoring from monthly to once every ten days.

Water samples are collected using a Nansen bottle at the surface, 1.5 m below the surface and at 3 m depths from each station at the same time the vertical plankton hauls are made. Water samples from the different depths are kept in separate bottles.

The water samples collected are brought to the laboratory and after determining the salinity and pH are treated with formaldehyde in the same manner as the plankton samples. If no *Pyrodinium* is present in the plankton samples of a station, the water samples are not subjected to further examination. However, if the plankton samples indicate the presence of *Pyrodinium* then, well mixed 10 ml subsamples of the water samples taken at different depths are centrifuged and subjected to detailed *Pyrodinium* counts. During the 1988 red tide event, other than for one water sample, no *Pyrodinium* was encountered although the plankton haul samples at times showed counts of more than 2,000 cells/0.01 ml. It is very likely that the *Pyrodinium* population could have been deeper than the maximum depth of 3 m down to which water samples were taken. In view of this possibility the sampling depths have been changed to include the surface, middle and bottom of the water column.

Stomach Content Analysis

The stomach content analysis of planktivorous fish have been included in the red tide monitoring program since the discovery that such fish could be implicated in PSP. On 4 December 1987 *Sardinella* spp. imported from East Malaysia were confiscated from markets and destroyed after discovering that over 80% of the stomach contents were composed of *Pyrodinium* cells. Subsequent PSP toxin analysis indicated a toxin level in excess of 400 µg/100 g of flesh for the stomach and associated organs. The flesh of the affected fish was, however, negative for PSP toxins.

The target species for stomach content analysis for *Pyrodinium* cells are planktivorous fish such as *Sardinella* and *Rastrelliger* species. Locally caught planktivorous fish selected randomly from markets are subjected to screening if *Pyrodinium* is detected in the coastal waters during normal monitoring. Imported fish are also subjected to such screening and particularly so, if red tides are reported from the country of import. Fish found to contain *Pyrodinium* are subjected to PSP toxin bioassay.

Monitoring PSP Toxins

The presence of PSP toxins in shellfish and fish is a direct effect of *Pyrodinium* red tides and their content in shellfish and fish has been universally used as a criterion to determine the severity of a red tide. In Brunei Darussalam, the standard mouse bioassay (Horwitz 1970) has been used for the toxicological determination of PSP. During the 1976 red tide, fish and shellfish samples were assayed initially by the Department of Pharmacology, University of Singapore, and then by the Brunei Medical Department in a mouse bioassay unit especially established for the purpose (Beales 1976). For the 1980 occurrence toxicological tests were carried out by the Department of Fisheries. Since then, samples have been sent to the Primary Production Department, Singapore, for toxicological investigations.

During "non-red tide" periods samples of the green mussel *Perna viridis* from the Gahasa Marine Aquaculture Farming Company (GAMAFCO) and from the aquaculture raft of the Department of Fisheries are sent to Singapore every month for mouse bioassay. During suspected red tides when *Pyrodinium* is detected in the water or in the stomach contents of target planktivorous fish, samples are sent every 10 days.

Recently a high-pressure liquid chromatography (HPLC) system has been installed in the Department of Fisheries specifically for assaying PSP toxins. A mouse colony is also being re-established to serve as a standby in case of need.

Red Tide Action Plan

In addition to the regular red tide monitoring and surveillance programs carried out in Brunei Darussalam, a Red Tide Action Plan has been developed which incorporates past experiences and field-tested procedures that have been beneficial in mitigating the impacts of red tides.

Objective

The objective of the Red Tide Action Plan is to provide timely and adequate responses to safeguard public health and to minimize economic losses during *Pyrodinium* red tide occurrences.

It establishes the procedures and responsibilities so that the response time can be reduced to a minimum during red tide occurrences and aims to:

- Safeguard public health against PSP due to red tides.
- Provide advance warning of impending red tides by regular monitoring.
- Reduce the response time in case of red tide occurrences.
- Provide accurate information to the public.
- Build up public confidence in the ability of the authorities to mitigate the effects of red tides.

Organizational Structure of the Red Tide Action Plan

Fig. 2 summarizes the organizational structure of the Red Tide Action Plan as given by Matdanan et al. (1988). The National Red Tide Response Team (NRTRT) made up of Senior Officers of the Department of Fisheries, Department of Medical and Health Services and the Municipal Board will be the decisionmaking and directing body. The Department of Fisheries will be responsible for executing the Red Tide Monitoring and Surveillance Program through the Environment Unit of its Marine Fisheries Section. Although the Department of Medical and Health Services will be directly involved in the safety of public health, the Department of Fisheries and the Municipal Board play prime roles in the red tide related public health problems. The former due to its overall jurisdiction over fish and fishing and the latter due to its control over markets and marketing. The Department of Broadcasting and Information plays a role in the dissemination of information on red tides through Radio and Television Brunei as well as other government publications such as the weekly news bulletin "Pelita Brunei".

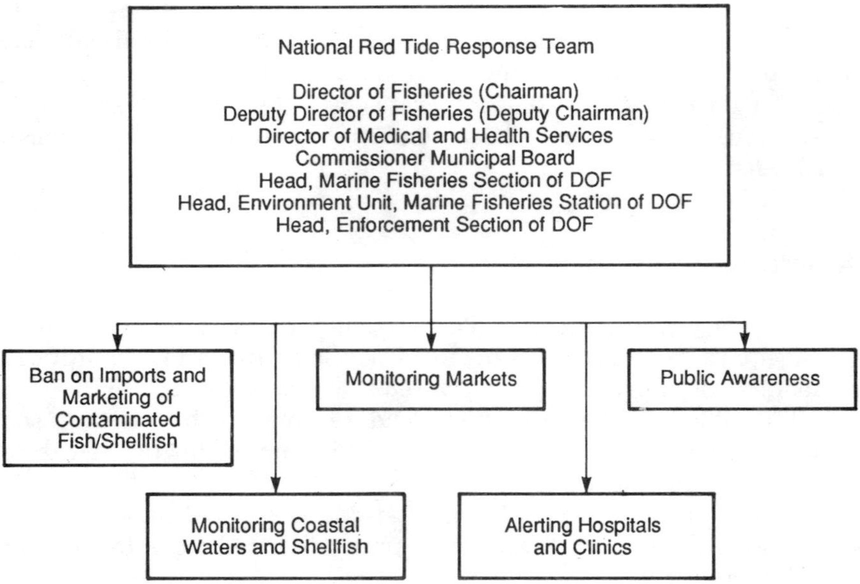

Fig. 2. Organizational structure of the red tide action plan.

Procedures of the Red Tide Action Plan

The general procedures that need to be adopted in case of a red tide occurrence are summarized in Fig. 3 and involve the following phases described in detail by Matdanan et al. (1988):

- Confirmation of the red tide occurrences.
- Steps to be taken on the confirmation of the red tide occurrences.
- End of a red tide occurrence.
- Review of the Red Tide Action Plan.

In addition to the routine monitoring and surveillance carried out by the Environment Unit (EU), Marine Fisheries Section of the Department of Fisheries, any information on actual or suspected red tides should be immediately conveyed to the Chairman/Deputy Chairman of the NRTRT. The Chairman/Deputy Chairman will alert all other members of the NRTRT. The following actions will then be carried out:

PHASE I - CONFIRMATION OF RED TIDE OCCURRENCE

a) The EU will proceed immediately to confirm the report and provide feedback to the NRTRT.

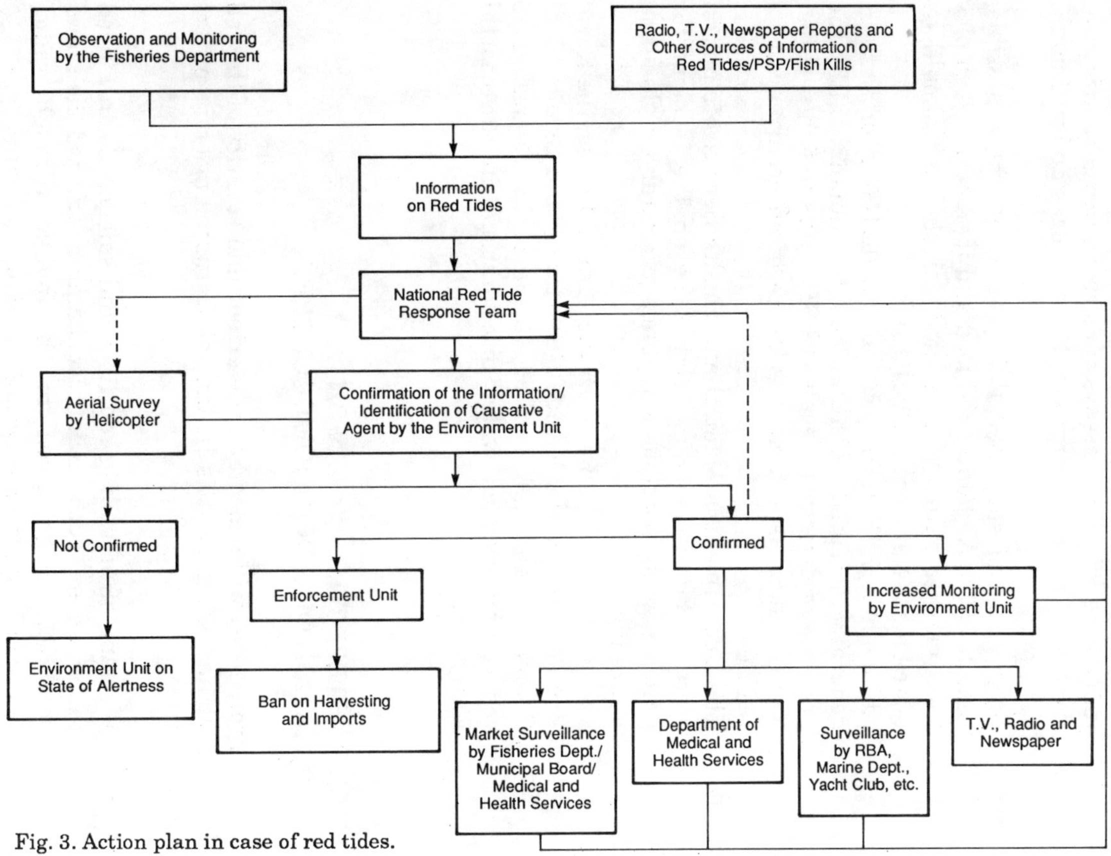

Fig. 3. Action plan in case of red tides.

b) Reports of water discoloration in the coastal waters outside the range of small boats to be investigated by helicopter.

c) Detection of unusual numbers of *Pyrodinium bahamense* var. *compressum* in vertical plankton haul samples or in the gut contents of planktivorous fish to be followed by immediate PSP toxin assays of cultured *Perna viridis* from the GAMAFCO farm, Serasa, Pelong Rocks and in Anduki sand pits.

d) In the case of suspected PSP poisoning; where possible, samples of cooked/uncooked food and if traceable, samples from market outlet of suspected fish/shellfish should be tested for the presence of PSP toxins.

e) Although a PSP toxin level of 80 μg/100 g of flesh is generally accepted as hazardous to human health, the presence of detectable levels of PSP toxins in fish/shellfish will be taken as sufficient evidence of an occurrence of a red tide.

f) The Department of Medical and Health Services to put on alert all hospitals and clinics to be on the look-out for patients with possible symptoms of Paralytic Shellfish Poisoning.

g) The Air-Wing of the Royal Brunei Armed Forces, the Royal Brunei Airlines, the Marine Department and Brunei Shell Company Sendirian Berhad (BSP) to be requested to report any unusual water discolorations in the coastal waters.

PHASE II - STEPS TO BE TAKEN ON CONFIRMATION OF A RED TIDE OCCURRENCE

The procedures will depend on the extent and severity of the red tide incident and whether it is in Brunei Darussalam or in others parts of Borneo.

a) Warning of prevailing red tides and safeguards that need to be taken, such as not consuming affected fish and shellfish, to be given to the public through radio, television, newspapers, posters and notice boards.

b) The Enforcement Section of the Department of Fisheries to ban the harvest/imports of molluscan shellfish and if necessary of planktivorous fish and other affected organisms.

c) The Department of Medical and Health Services, the Department of Fisheries and the Municipal Board to keep surveillance over markets and other sales outlets to see that no banned fish/shellfish are sold.

d) The EUC to reduce the interval of monitoring for:
   i) red tide organisms in the coastal waters to once a week instead of the usual once a fortnight procedure;
   ii) PSP toxins to once in ten days instead of monthly;

e) Increase monitoring stations to cover Temburong, Tutong and Belait Districts.

f) The EUC to maintain close contact with the Fisheries Department of the East Malaysian States of Sabah and Serawak with respect to the red tide situation.

PHASE III - END OF A RED TIDE OCCURRENCE

Recurrent negative results for PSP toxins in test shellfish for one month together with only the occasional occurrence of a few cells of *P. bahamense* var. *compressum* in plankton haul samples over the same period of time can be taken to signal the end of a red tide. The following action is then to be taken:

a) The NRTRT should make a public announcement that the red tide alert is removed. The same media utilized to warn the public of a red tide occurrence should be used for the purpose.

b) All posters and boards put up to warn the public regarding a red tide occurrence should be removed unless they were put up for the purposes of public education or awareness on red tides. Particular attention should be given to boards banning the harvesting of shellfish from specific locations.

c) EU to revert back to normal monitoring and surveillance procedures.

PHASE IV - REVIEW OF THE RED TIDE ACTION PLAN

It would be necessary to review the Red Tide Action Plan after the occurrence of a red tide incident to strengthen areas of weakness. Further, at least an annual review needs to be made to maintain the accuracy of emergency contact addresses and telephone numbers. Updating of detection techniques for PSP toxins and the red tide organisms would be necessary because of the rapid progress being made in these areas.

The Environment Unit, Marine Fisheries Section of the Fisheries Department, in addition to executing the Red Tide Monitoring and Surveillance Program will be responsible for logging the events during a red tide occurrence. It will also be the repository for all details of procedures relating to the Red Tide Action Plan. This would include in addition to all emergency contact telephone numbers and addresses, the details of sampling and analytical procedures as well as drafts of memoranda, minutes and public announcements, etc.

In Brunei Darussalam fish and shellfish recording positive PSP toxins are taken as being undesirable for consumption. The uncertainties associated with PSP toxin levels capable of causing human health problems including death and the lack of information on human safety threshold levels for *Pyrodinium* PSP toxins have warranted such a decision.

References

Beales, R.W. 1976. A red tide in Brunei's coastal waters. Brunei Mus. J. 3(4): 167-182.

Horwitz, W. 1970. Official methods of analysis of the Association of Official Analytical Chemists. AOAC. Washington, DC.

Matdanan Haji Jaafar and S. Selvanathan. 1984. Occurrences of red tide in Brunei Darussalam and methods of monitoring and surveillance, p. 17-27. *In* A.W. White, M. Anraku and K.-K. Hooi (eds.) Toxic red tides and shellfish toxicity in Southeast Asia. Southeast Asian Fisheries Development Center, Singapore.

Matdanan Haji Jaafar, M.W.R.N. De Silva, P.H.Y. Sharifuddin and Siti Amin. 1988. A Red Tide Action Plan for Brunei Darussalam. Paper presented at the ASEAN-US Coastal Resources Management Project Technical Workshop on Integrated Tropical Coastal Area Management. National University of Singapore, Singapore.

Management of Red Tides in Sabah, Malaysia

JOSEPH TUNG SANG WONG
THIAN MING TING
Department of Fisheries
Sabah, Malaysia

WONG, JOSEPH TUNG SANG and THIAN MING TING. 1989. Management of red tides in Sabah, Malaysia, p. 135-139. *In* G.M. Hallegraeff and J.L. Maclean (eds.) Biology, epidemiology and management of *Pyrodinium* red tides. ICLARM Conference Proceedings 21, 286 p. Fisheries Department, Ministry of Development, Brunei Darussalam, and International Center for Living Aquatic Resources Management, Manila, Philippines.

Abstract

In response to the occasional incidence of paralytic shellfish poisoning in Sabah, the Fisheries Department undertakes a number of precautionary measures, including fortnightly monitoring of toxin levels in shellfish and of plankton samples for *Pyrodinium bahamense* var. *compressum,* the causative dinoflagellate; public warnings by word-of-mouth, circulars and posters and then by broadcasts; education campaigns and advice on safe levels of shellfish consumption. Other government agencies and private industry also perform various roles in reducing the risk of PSP incidence.

Introduction

Outbreaks of red tide and paralytic shellfish poisoning (PSP) from consumption of toxic shellfish have occurred from time to time in certain parts along the west coast of Sabah, Malaysia (Ting and Wong, this vol.) caused by the dinoflagellate *Pyrodinium bahamense* var. *compressum.* Because of the danger the red tide outbreaks pose to the public health as well as the livelihood of the fishermen, the government is taking various measures to reduce poisoning risks including such precautionary measures as monitoring, public warnings, and educational and other measures.

Precautionary Measures

Red Tide Monitoring

In Sabah, shellfish often become toxic without any visible warning such as a plankton blooms; toxic shellfish cannot be distinguished by taste, appearance or smell. Therefore, shellfish toxicity and plankton in the sea have to be constantly monitored to control and minimize health risks.

Red tide monitoring essentially involves regular plankton sampling and checking of toxin levels in shellfish in key sampling locations. In Sabah, key sampling areas and high-risk zones have been identified, most of which are in the west coast. Shellfish and plankton samples are collected fortnightly from these chosen locations. The collection frequency may be increased depending on the prevailing situation such as whether there is an outbreak of red tide or PSP in the area. These samples are sent as soon as possible to the Fisheries Research Center in Kota Kinabalu and where delay is anticipated the samples are suitably preserved with ice or formalin. At the Research Center the plankton samples are filtered and examined and dominant species of plankton identified and counted. Toxicity in shellfish is determined by mousebioassay according to the AOAC official method. The Research Center handles around 600-800 shellfish and a similar number of water samples annually. The results so obtained are relayed to the fisheries headquarters in Kota Kinabalu and to the district fisheries office concerned for further action.

Public Warning

Whenever toxin score reaches or exceeds the danger level of 80 µg/100 g (based on Canadian Standard) or when the number of *Pyrodinium* cells reaches a few thousand per liter, a public warning is issued to the area concerned. The warning must reach the target groups in the shortest time possible and, as experience has shown, the fastest and the most effective means is spreading the news by word-of-mouth, which means going into the affected area and give warning directly to the district officer, village chief, community leaders and to as many kampung (village) folks as possible. This will be followed by a warning circular and putting-up of warning posters at vantage points in the affected location. In the meantime, the Medical Department is informed of the situation.

The warning circular is normally worded as follows:

1. With immediate effect cease picking, selling or eating all types of shellfish, including snails, until further notice.
2. All fish, shrimp, crabs and lobsters are safe to be eaten on condition that their gills and guts are removed and the fish, shrimp, etc. are then thoroughly washed with clean running water.
3. In the event of suspected poisoning, seek medical help immediately.

The circular is addressed to the village chiefs, the District Officer, People's Development Leader and the local vigilante corps who will in turn relay the message to more people in the affected areas.

If the situation warrants wider publicity such as when red tide or PSP prolongs or spreads over a wider area then the Fisheries Department, with the consent of the Ministry of Agriculture and Fisheries, may issue a press release through the Department of Information. Radio interviews, broadcast direct, may also be conducted in order to reach the widest possible audience in a very short time.

It has been proven that spreading the news verbally, followed by warning circulars and posters can best alert the public without causing any undesirable side effects such as a public panic. On the other hand, indiscriminate reports can create havoc among the general public. Such reporting can be very misleading and inevitably it scares off people from eating even wholesome seafood, to the detriment of the fishermen and fishmongers alike. As the public adopt an eat-safe attitude, the prices of fish may plummet and the fishermen may suffer loss of income. As a result, the Government may have to provide them assistance. The prices of other food items may shoot up, the Fisheries Department telephone lines may be jammed with inquiries from the anxious public, while exports of fish and other seafood may also be affected. This undesirable chain reaction is not benefiting anybody and can easily be avoided if the press is more careful and accurate in its reporting and reports only correct information from authorized sources.

Public Education

All monitoring, warning and other efforts carried out by the government related to red tide and PSP would be futile if the public were not receptive to the Government education effort on the basic facts about red tide. There is still a large cross-section of people who are not well informed on the subject of red tide. The adage, "where ignorance is bliss, it is folly to be wise", does not quite apply here, for ignorance creates panic and worse still renders people gullible to myths and half-truths. The end result is that the public becomes polarized in its opinions. To one group the word "red tide" spells fear and anxiety and such people abstain from all types of seafood. The other group may choose to deliberately ignore the warning and eat as they please. If the people were well informed of the risks and danger of PSP and other vital information they would act more calmly and follow and appreciate the advice from the relevant government departments. Very often those who suffer from PSP are those who deliberately choose to ignore warning and advice given by the relevant authorities. There is evidence that some of them resort to vandalism by destroying warning posters put up by the Fisheries and the Medical Departments.

Thus, public education is as vital as the red tide monitoring itself. The Fisheries Department is sparing no effort in trying to educate the public, especially people living in the high risk areas along the west coast of Sabah. Besides the radio interviews broadcast direct at prime time and holding talks in the villages telling the people about red tide and PSP in simple language, the Fisheries Department has held numerous red tide awareness, campaigns cum exhibitions in conjunction with the State festivities and also has printed leaflets on some important aspects of red tide, including advice and precautionary measures. The Fisheries Department is also planning to hold a public forum on red tide in the very near future in order to enhance further public awareness of PSP.

Other Measures

The Fisheries Department is not taking any chances when it comes to eating shellfish harvested from previously affected areas. During the so-called safe season when no or very few *Pyrodinium* cells are detected in the waters or when no toxin is found in shellfish samples from a particular area, the people are nevertheless strongly advised to eat at each meal only a small amount of shellfish collected from the previously affected area. **Under no circumstances should a large amount of shellfish be consumed at one single meal.** This is because of the possibility of the shellfish becoming toxic suddenly without any visible warning or before being detected by the relevant authorities.

Parents are also strongly urged to prevent their children from eating shellfish even during "safe" periods as children are more susceptible than adults to paralytic shellfish toxins.

Roles of Government Agencies

The bulk of the monitoring work and public warning is carried by the Fisheries Department but the cooperation of other relevant government agencies is essential, especially that of the Medical and the Information Departments which have been extremely helpful. The Information Department, by the use of its mobile unit, has assisted in the dissemination of red tide warnings and other related information. The Medical Department has assisted in some of the shellfish collection in remote areas, in reporting and investigating PSP cases and also in enforcing the shellfish ban during PSP outbreaks.

Civil Aviation has also done its part by reporting to the Fisheries Department any red tide patches its pilots may spot during flights.

The Fisheries Department has also appealed to the fishermen, Marine Police, Navy and offshore oil companies to report to the nearest fisheries office sightings of any red patches in the sea so that a fast and appropriate follow-up action can be carried out to assess the situation.

Management of Toxic Red Tides in the Philippines

CIELITO L. GONZALES
Bureau of Fisheries and Aquatic Resources
860 Quezon Avenue, Quezon City
Philippines

GONZALES, C.E. 1989. Management of toxic red tides in the Philippines, p. 141-147. *In* G.M. Hallegraeff and J.L. Maclean (eds.) Biology, epidemiology and management of *Pyrodinium* red tides. ICLARM Conference Proceedings 21, 286 p. Fisheries Department, Ministry of Development, Brunei Darussalam, and International Center for Living Aquatic Resources Management, Manila, Philippines.

Abstract

Since its first occurrence in 1983, toxic red tides, caused by *Pyrodinium bahamense* var. *compressum,* have become one of the major health and fishery problems in the Philippines. The intoxication resulting from the consumption of red tide-contaminated shellfish and the inaccurate reporting about red tide in the local mass media resulted in great economic loss to the fishing industry.

The methods adopted by the Philippine government in the management of the red tide problems are presented here.

Introduction

Toxic red tides have been one of the major health and fishery problems in the Philippines since it first occurred in Samar in 1983. Since then, the phenomenon continued to adversely affect not only the fishing industry, as a result of the red tide scare, but also the population of the affected areas due to the paralytic shellfish poisoning (PSP) it causes to those who eat contaminated shellfish. Most people became scared and avoided eating marine fisheries products, including those caught from other fishing grounds not affected by the toxic red tides. This resulted in great economic loss to the fishing industry particularly during the 1988 red tide occurrence in Manila Bay which, in the early stage of the toxic dinoflagellate bloom, caused approximately P17 million (P21 = US$1.0 in 1988) loss in four days to commercial fisheries alone (Robles 1988).

Since 1983, there have already been seven major outbreaks in the country: two in Samar-Leyte (1987 and 1988), one in Zambales (1987), and one each in Manila Bay (1988), Negros Occidental (1988), Capiz (1988) and Cebu (1989). To date, a total of 1,200 PSP cases were reported from these areas of which 42 persons died (Estudillo and Gonzales 1984; Gonzales et al. 1989; Pastor et al. 1989). Shellfish, particularly the green mussel *(Perna viridis),* were implicated in the majority of the PSP cases.

Starting in 1984, the government implemented a monitoring program for all red tide-affected areas to detect a bloom at its early stage in order to minimize, if not totally eliminate, its effect on public health.

The methods of water and shellfish monitoring, public education campaign, medical management of PSP cases, and other government interventions during the toxic red tide occurrences in the Philippines since 1983 are presented in this paper.

Strategies

Red Tide Monitoring

Plankton and shellfish samples are collected fortnightly from areas with histories of *Pyrodinium* blooms.

Plankton samples are collected by vertical haul with a 40-µm mesh plankton net from near the bottom to the sea surface. The samples are preserved either in Lugol solution or in 10% formalin solution, and brought to the Central Laboratory in Metro Manila for qualitative and quantitative analyses. Counts of *Pyrodinium* cells are made from a 1-ml aliquot placed in a Sedgewick-Rafter counting chamber. These counts are the basis of the numerical estimates of the plankton population, taking into consideration the mouth diameter of the net, sampling depth, and volume of plankton sample.

During red tide blooms, aerial surveillance on board helicopters and light aircraft of the Philippine Air Force and the Department of Agriculture (DA) are carried out to determine the extent of the bloom and the movement of the visible red tide. Based on the information gathered from these aerial observations, the residents in affected areas are alerted.

In Bataan and Capiz provinces, Red Tide Task Forces, composed of local officials and personnel from the government health and agriculture agencies, were formed to coordinate red tide monitoring activities and to inform the public of the latest development on the red tide situation in the area.

Stomach content examinations of shellfish samples from shellfish farms, public markets and fish landing sites are made periodically. Shellfish samples for toxicity tests are shucked in the field, frozen and transported to the Bureau of Food and Drugs (BFAD) in Metro Manila where the level of toxicity of the samples is determined by the standard mouse bioassay tests according to the Association of Official Analytical Chemists (AOAC) method (Horwitz 1980).

Other hydrobiological parameters such as water temperature, salinity, pH, dissolved oxygen, phosphate-phosphorus, and cyst density are determined once a month during neap tides to minimize the tidal effect on the water samples.

Meteorological conditions such as amount of rainfall, wind force and wind direction are taken from the records of the Philippine Atmospheric, Geophysical and Astronomical Services Administration (PAGASA).

Information Dissemination

Public announcements of red tide alert through print and broadcast media are made whenever the toxicity in shellfish exceeds the regulatory limit of 80 μg toxin per 100 g of shellfish meat. During such situations, the public is advised to refrain from eating any kind of shellfish from red tide-infested waters. To avoid misconception and its effect on the sale of other marine fishery products, the public is also informed that fish and other invertebrates caught even from red tide areas are safe for human consumption, provided that they are fresh, eviscerated and washed thoroughly before cooking.

Public meetings and seminars are organized, especially in the fishing villages, to inform the fishermen and the public properly on the current red tide situation. In addition, village criers are sent out to issue warnings to the villagers that toxic red tide is present in their area and eating shellfish should, therefore, be avoided.

Primers on red tide are distributed and audio-visual presentations of information on red tide are shown to the public through local televisions and the "Agri Vans" of the Department of Agriculture.

Prompted by inaccurate reporting about red tide in the local print and broadcast media, a Red Tide Committee, composed of personnel from the Department of Agriculture (DA) and the Department of Health (DOH), was formed during the toxic red tide occurrence in Manila Bay in 1988. The committee evaluated current data on the red tide occurrence and based on the results of the data evaluation, one-fourth page Red Tide Updates were published weekly for eleven weeks by the two agencies in several newspapers in order that the public would be properly guided on which fisheries products are safe for human consumption and which should be avoided. The move, apparently, was effective because after the first update was issued, the PSP epidemic was contained.

In order to develop capabilities in the regional offices, the DA, through the Bureau of Fisheries and Aquatic Resources and the Agricultural Training Institute, conducted in 1987 a national training course on toxic red tide detection, identification and monitoring. Sixty-six technical personnel, representing all regional offices of the DA throughout the country attended the training course. They now form the core groups that monitor their respective areas for the presence of toxic dinoflagellates in the water and PSP toxins in shellfishes, so as to be able to respond quickly should a threat of a toxic red tide occurrence become imminent.

Regulation

As soon as the toxin level in shellfishes exceeds the regulatory limit of 80 µg toxin per 100 g shellfish meat, the DA imposes a temporary ban on the harvesting, marketing and transporting of all kinds of marine shellfish from the red tide-contaminated waters. The issuance of auxillary invoices, a requirement in transporting fishery products from one place to another, is suspended to prevent the movement of contaminated shellfish to non-affected areas. Checkpoints are also established at strategic locations, such as piers, airports and bus routes, to ensure that no contaminated shellfish are moved out of the red tide areas.

Medical Management of PSP Cases

There is no specific antidote for the PSP toxin, and the manner of treatment is largely symptomatic (Halstead 1965). In the Philippines, it is suggested that the first thing to consider in PSP management is to empty the stomach of the victim of the toxic material as quickly as possible. This may be done by giving the patient an oral emetic or by simply inserting a finger in his/her throat to induce vomiting. In as much as the toxin is water soluble, it is also recommended that the victim be given copious amount of water to induce urination and minimize the absorption of the toxin by the gastro-intestinal tract.

In the hospital, the DOH and the Philippine General Hospital follow a protocol on the management of PSP cases. Patients not exhibiting respiratory distress are given nasogastric infusion of sodium bicarbonate in water (0.5 g/kg body weight), activated charcoal lavage (30 g in 100 ml water as slurry for adults or 10 g for less than 2-year old children, 20 g for children between 3 and 7 years and 30 g for children of more than 8 years) after 5 minutes, reinfusion of the same amount of activated charcoal, sodium sulfate catharsis (30 g in 250 ml water or 0.25 g/kg body weight in water), and 1-2 meq/kg-body-weight intravenous (IV) infusion of sodium bicarbonate for 24 hours.

When respiratory distress is present, endotracheal entubation is performed on the patient and artificial ventilation is given. The patient is then admitted to the intensive care unit and if respiratory paralysis is manifested, neostigmin, at 1-3 mg IV or 0.02 mg/kg body weight IV every four hours, and atropine at 0.01 mg/kg body weight IV, are given. Physostigmin or endrophonium may be used in place of neostigmin (Anon. 1988).

The universal antidote used in Samar for all kinds of poisoning contracted through the oral route, is pure coconut milk. During the first toxic dinoflagellate bloom in 1983, the use of coconut milk as an antidote for the red tide toxin was tried. Based on the records of the Catbalogan Integrated Provincial Hospital in Samar, more than 80% of the patients given an oral dose of the local antidote recovered. The levels of toxicity of the contaminated shellfish ingested by the victims could not be determined to make a more meaningful evaluation of the efficacy of coconut milk as an antidote. However, preliminary results of experiments conducted by the Department of Health Regional Office No. VIII and the Southeast Asian Fisheries Development Center (Gacutan 1986) suggest that the local cure for PSP is really effective. It should be emphasized, however, that coconut milk, or any oral drugs, should not be given to patients starting to show symptoms of dysphagia and respiratory failure due to the risk of asphyxia.

Government Assistance Program

As a result of the red tide scare, the income of 44% of fishermen and fish vendors from some fishing towns around Manila Bay had decreased substantially while 42% earned nothing at all (Batnag 1989).

Due to economic hardship suffered by fishermen during red tide occurrences, the government, through non-government organizations and fishermen's cooperatives, granted the red tide-affected artisanal fishermen emergency loans amounting to P3,000 to 5,000 (US$142 to 238) each, payable within a period of one year with an interest rate ranging from 5 to 12% per annum (F. Matienzo, pers. comm.). The proceeds of loan payment go to a community revolving fund and are lent again to other beneficiaries.

Two months after the first PSP cases were reported from Manila Bay, the red tide toxins in shellfish became zero, except in a small area in Bataan. This delayed the lifting of the shellfish ban, but due to the financial difficulties the shellfish farmers were experiencing, the DA decided in December 1988 to harvest the remaining contaminated green mussels in that area, which paved the way for the early lifting of the ban.

Future Plans

The Philippine government recognizes the importance of the early detection of toxic red tide occurrence as a key to the prevention of PSP epidemics. This could be made possible by collecting shellfish samples at a regular interval, e.g., weekly, and subjecting them to mouse bioassay. Presently, however, shellfish samples collected from monitoring sites still have to be sent to the BFAD in Metro Manila for the bioassay. It takes a few days to more than a week before the results are known because of the workload of the BFAD laboratory. Efforts are exerted to develop the regional and provincial capabilities of the DA and the DOH on bioassay techniques to at least determine the presence or absence of the toxin in shellfish samples. If PSP toxins were detected from the sample, the necessary warning may be issued while waiting for the results of the confirmatory tests to be performed on the samples by BFAD. This way, the government may be able to respond quickly should there be a recurrence of the toxic red tide.

References

Anon. 1988. Paralytic shellfish poisoning. Seminar on Red Tide. Philippine General Hospital, Manila. 7 p. mimeo.

Batnag, D. 1989. Red tide has changed lives of small fishermen. Manila Chronicle, March 21, 1989: 1.

Estudillo, R.A. and C.L. Gonzales. 1984. Red tides and paralytic shellfish poisoning in the Philippines, p. 52-79. *In* A.W. White, M. Anraku and K.K. Hooi (eds.) Toxic red tides and shellfish toxicity in Southeast Asia. Southeast Asian Fisheries Development Center, Singapore, and International Development Research Centre, Ottawa, Canada.

Gacutan, R.Q. 1986. Effects of coconut milk and brown sugar on crude toxins from mussels exposed to *Pyrodinium bahamense* var. *compressa,* p. 311-313. *In* J.L. Maclean, L.B. Dizon and L.V. Hosillos (eds.) The First Asian Fisheries Forum. Asian Fisheries Society, Manila, Philippines.

Gonzales, C.L., J.A. Ordoñez and A.M. Maala. 1989. Red tide: the Philippine experience, p. 45-48. *In* T. Okaichi, D.M. Anderson and T. Nemoto (eds.) Red tides, biology, environmental science and toxicology. Elsevier, New York.

Halstead, B.W. 1965. Poisonous and venomous marine animals of the World. Vol. 1. Government Printing Office, Washington, DC, USA.

Horwitz, W. 1980. Paralytic shellfish poison, biological method. *In* Official methods and analyses. 13th ed. Association of Official Analytical Chemists, Washington, DC, USA.

Pastor, N.I.S., I.L. Gopez, M.C. Quizon, N.B. Bautista, Jr., F.M. White, V.S. Lofranco, R.M.S. Benabaye, M.E. White and M.M. Dayrit. 1989. Epidemics of paralytic shellfish poisoning in the Philippines, 1988-1989, lessons in science and public policy. 14 p. mimeo.

Robles, A.C.. 1988. Red tide: ignorance takes its toll. Manila Chronicle, September 25, 1988: 9.

Management Approaches to Red Tides in Papua New Guinea

K. RIRORIRO
L.D. SIMS
Department of Agriculture and Livestock
National Veterinary Laboratory
Kila Kila, Port Moresby
Papua New Guinea

RIRORIRO, K. and L.D. SIMS. 1989. Management approaches to red tides in Papua New Guinea, p. 149-151. *In* G.M. Hallegraeff and J.L. Maclean (eds.) Biology, epidemiology and management of *Pyrodinium* red tides. ICLARM Conference Proceedings 21, 286 p. Fisheries Department, Ministry of Development, Brunei Darussalam, and International Center for Living Aquatic Resources Management, Manila, Philippines.

Abstract

No confirmed cases of toxic *Pyrodinium* blooms have been reported in Papua New Guinea during the past 18 years, although it is highly likely that unreported incidents occurred. Past instances were treated on an *ad hoc* basis. A National Disasters Emergency Plan has recently been introduced and could be extended to include red tides.

Introduction

The first documented report of red tides and associated paralytic shellfish poisoning in Papua New Guinea occurred in 1961 (Rapson 1968) although the condition had undoubtedly been present in the country for many years prior to this outbreak (Maclean 1973). Further outbreaks up to 1973 have also been documented (Maclean 1973; Rhodes et al. 1975) but records from that time onwards are rather sketchy. The last reported "red tide" was in 1986 from Milne Bay Province but no disease was associated with it (A. Richards, pers.

comm.) Medical statistics are also incomplete and it is likely that cases of paralytic shellfish poisoning have occurred (apart from those reported by Maclean (1973) and Rhodes et al. (1975)) but remained undiagnosed or unreported.

Papua New Guinea relies heavily on its marine resources for subsistence, for access to the cash economy and for export earnings. Approximately 400,000 people live in coastal areas (Anon. 1989) and many of these reside in remote villages where they have limited access to government services and poor road and communication networks. These factors are important constraints to the development and implementation of management plans for red tides.

Present Situation

Regular surveillance for red tides is not undertaken. A formal action plan is not in existence and each report in the past has been dealt with on an individual basis. The National Department of Fisheries and Marine Resources is the usual point of first contact for reports of red tides. Reports would usually come from a fisheries officer based in the affected region. Given the focal nature of red tides and the limited number of fisheries staff there is scope for red tides to go unreported. Following the recognition of a red tide a warning is issued to villagers in the affected region by way of the National Broadcasting Commission's radio services and the national press. Extension material in the form of an article in the government magazine Harvest was prepared in the early 1970s. Pamphlets or other handout material are not currently available.

Initial diagnosis of the problem could come from staff of the Department of Health. However, primary health care at the village level is basic and health workers are unlikely to recognize the signs of the disease unless forewarned of its potential occurrence.

Laboratory services to confirm the identification of *Pyrodinium* would be provided by the National Veterinary Laboratory, which is part of the Department of Agriculture and Livestock. This laboratory performs microbiological and toxicological testing on seafoods (especially those destined for the export market) on behalf of the Department of Fisheries and Marine Resources.

The National Government has recently introduced a National Disasters Emergency Plan which could be implemented or modified if red tides and associated diseases were reported and considered sufficiently severe for concerted action to be taken. At present they have not been considered for inclusion in this plan.

There are some villages and areas where the link between red tides and food poisoning is well recognized and tradition dictates that shellfish are not consumed for several months after the occurrence of red tides. The extent to which this tradition is practised is not known but it probably contributes significantly to reducing the morbidity associated with paralytic shellfish poisoning.

References

Anon. 1989. Fisheries sector review, Papua New Guinea. United Nations Development Program, Washington.

Maclean, J.L. 1973. Red tide and paralytic shellfish poisoning in Papua New Guinea. Papua New Guinea Agric. J. 24(4): 131-138.

Rapson, A.M. 1968. Ichthyosarcotoxism (fish poisoning) in New Guinea. SPC/ICHT/WP.10. South Pacific Commission, Noumea, New Caledonia.

Rhodes, F.A., C.G. Mills and K. Popei. 1975. Paralytic shellfish poisoning in Papua New Guinea. Papua New Guinea Med. J. 18: 197-202.

Management of Red Tides and Paralytic Shellfish Poisoning in Guatemala

F. ROSALES-LOESSENER
DITEPESCA
Guatemala

ROSALES-LOESSENER, F. 1989. Management of red tides and paralytic shellfish poisoning in Guatemala, p. 153-154. *In* G.M. Hallegraeff and J.L. Maclean (eds.) Biology, epidemiology and management of *Pyrodinium* red tides. ICLARM Conference Proceedings 21, 286 p. Fisheries Department, Ministry of Development, Brunei Darussalam, and International Center for Living Aquatic Resources Management, Manila, Philippines.

Abstract

A modest shellfish toxicity monitoring program was set up in Guatemala following the 1987 outbreak of paralytic shellfish poisoning (PSP). Shellfish positive for PSP toxins have been found in 1987, 1988 and 1989. Problems in the program are outlined.

Introduction

After the July 1987 PSP outbreak experiment in Guatemala, it was necessary for the government public health and fisheries authorities to establish an action plan in order to protect the public and prevent future human tragedies and economic losses caused by red tides and PSP.

Therefore, the Red Tide and PSP Commission was created, consisting of different public institutions from the Public Health, Defense and Agriculture Ministries. The main task of this Commission was to elaborate a proposal for a "Red tide and PSP monitoring program" in line with national needs and capabilities. The Commission was assisted by Dr. Sherwood Hall, who gave advice and guidelines on the basic technical criteria to be taken into account in implementing and executing the monitoring program.

Due to the constraints and lack of personnel and logistic support, the monitoring program was modest and simple. The main objective was to prevent cases of PSP in the Pacific coast through timely detection of PSP levels in bivalve samples assayed.

The Pacific coast of Guatemala is about 250 km long. Three sampling points were selected along the coast of Las Lilas; Iztapa and Champerico.

The shellfish samples were taken mainly from mangrove/estuarine channels which are located parallel to the open coastline. These mangrove ecosystems are the country's main source of commercial shellfish. There are no shellfish in the open coast except for the seasonal occurrence of the small donacid clam, *Amphichaena kindermanni,* which was the transvector of the 1987 PSP outbreak in Guatemala (Rosales-Loessener, this vol.).

Monitoring Program Results

From September to October 1987, 36 shellfish samples were collected and bioassayed. From these samples, 9 were found to be positive for PSP toxins, with levels ranging from 30 to 78 μg/100 g. All except one of the positive samples were from estuarine channels.

From March to December 1988, 89 shellfish samples were collected and bioassayed. Toxicity of these samples ranged from 20 to 51 μg/100 g. All samples were from the estuarine environment.

Twenty-five shellfish samples have been collected and bioassayed from January to May 1989. Two of them were found to be positive for PSP toxins with levels of 22 μg/100 g. These samples were also collected from estuarine environments.

Problems

The main problems encountered with the monitoring program are:

- Shellfish samples are not obtained from the ocean but from the inland estuarine channels where they are collected for local consumption and the domestic market.
- The increase in the toxicity level in the oceanic clams involved in the PSP outbreak in Guatemala two years ago was apparently very rapid. The clams became deadly toxic overnight.
- Responsibilities of the monitoring program are distributed among several government institutions which can create inefficiency at times.
- There is hardly any budget for the program. The government is trying to obtain funds through international cooperation programs.
- Lack of institutional personnel who are trained in red tides and PSP aspects.

Methods of Controlling *Pyrodinium* Red Tides

Moderator : T. Okaichi

Control of *Pyrodinium* blooms by chemical, biological and physical means was discussed.

Chemical Control

Recent experiences in Shido Bay, the Seto Inland Sea, Japan, were reported. Following strong requests from the local fishermen's association and with the cooperation of local government, 200 kg of sodium percarbonate (Na_2CO_3 3/2 H_2O_2) was sprinkled onto *Chattonella antiqua* red tide there in August 1987. The area treated was 2,000 m^2 with 2-2.5 m depth. Two hours after the application corresponding to 50 ppm, 90% of the *Chattonella* were eliminated.

Dilute formalin was another suggested control agent, at least in enclosed areas, such as fish or shrimp ponds.

Reservations were expressed as to the usefulness of this approach. Due to the vast scale of red tides, chemical control would be very expensive and potentially catastrophic to other organisms. Laboratory tests are needed first to investigate both short- and long-term effects of such chemicals on the marine environment.

Methods for control of red tide phytoplankton using chemical or biochemical agents are clearly experimental at present and thus should be pursued only by those countries with sufficient financial resources and economic justifications.

Biological Control

Tintinnids (e.g., *Favella*) have been observed to consume *Pyrodinium* in Sabah, but may not be effective control agents, since there is no evidence of high tintinnid grazing rates on *Pyrodinium*.

Physical Control

P. bahamense var. *bahamense* requires sunlight every day. As little as 24 hours in darkness leads to initiation of dying. Therefore, it might be feasible to "dust" the surfaces of well-defined coastal red tides with charcoal powder, to serve as an absorber of light. This layer should remain within the surface lens. The charcoal is otherwise innocuous and would eventually sink to bottom sediments.

The positive phototaxis of *P. bahamense* should make during hot, calm weather, a dense surface lens. Therefore, it is possible to use the techniques of oil slick skimming to filter the red tide slicks and thus physically remove them from the ecosystem.

Such methods would again be expensive given the size of *Pyrodinium* red tide blooms. Oil skimmers would be ineffective when blooms remain subsurface to any extent.

Another view was that coastal engineering works would be appropriate if the potential economic losses warrant them. For example, the phosphorescent Puerto Mosquito in Puerto Rico lost its blooms inadvertently when a new channel was cut through to the sea changing the flushing pattern.

"Anticipatory" Control

A more realistic investment of effort would be to "control" red tides by prudent management of the affected fish or shellfish resources. For example, by studying areas presently affected by toxic outbreaks, it should be possible to define the hydrographic, topographic, and cyst distributional patterns that are commonly associated with toxicity. Siting of new aquaculture facilities should then be guided by this knowledge so that areas are chosen where toxic blooms are unlikely. In the case of shellfish farms, which require abundant phytoplankton, sites with well-mixed water columns that would be more supportive of diatoms than dinoflagellates (which prefer stable, stratified conditions) should be selected. Conversely, fish farms should be located in areas where dense phytoplankton blooms of any type are not likely to occur. In such cases, control of red tides is "anticipatory" through management of the resources.

Monitoring Principles

Chairperson : G.M. Hallegraeff
Rapporteur : S. Blackburn

The toxic dinoflagellate *Pyrodinium* can contaminate shellfish and planktivorous fish, and when humans consumer these seafoods, this can result in paralytic shellfish poisoning (PSP) and, in extreme cases, death. The key monitoring principle is to survey *seafood* products for toxins before they are made available for human consumption.

Mouse Bioassay or HPLC?

To date, the only internationally accepted method for toxin testing is the mouse bioassay. The method is relatively east to perform and, despite inherent weaknesses such as variability (± 20% error) and delay between sample collection and analysis, has proved its value in Canada and the USA. The 80 µg/100 g quarantine level a very conservative value, set for convenience only and needs to be re-examined for *Pyrodinium*. For remote stations where a mouse colony is not available, altnerative qualitative assays using chicks chicks (Philippines), cats or flies (1-cm size) can be used. In some countries, animal assays are forbidden (Norway) or severely limited (Australia) and HPLC will be the method of choice. In Tasmania, for example, shellfish are routinely tested by HPLC and positive test results are confirmed by mouse bioassay before management decisions as to closure of shellfish farms can be made. In the Indo-West Pacific, dedicated HPLC equipment is now available in Brunei and the Philippines. Drawbacks of the HPLC method are its initial high expense, long development time and limited availability of suitable standards, although the US FDA has indicated that adequate standards will be produced in future.

How Many Stations? Water Samples or Seafood Samples?

Monitoring programs may be initially more extensive (and expensive) in order to identify key stations. Singapore maintains five stations in fish farm areas, Guatemala has two fortnightly stations, whereas the Philippines has problems in monitoring a larger number of remote islands. Tasmania has selected 1 or 2 key stations which always become toxic first and subsequently the sampling network can be expanded to 15 weekly or fortnightly stations. The choice of seafood product to be tested is also important. In Tasmania, the major commercial shellfish products are oysters but mussels (which become toxic earlier) are used for initial testing to provide an early warning signal. Similarly, Brunei is keeping green mussels in a fish farm area for testing purposes. While a lot of experience is available on testing (sessile) shellfish from well-defined areas, effective monitoring of (mobile) *planktivorous fish* from diverse and often distant areas is a daunting task and requires more research. Dinoflagellate monitoring on the basis of integrated water column samples (which requires a plankton net and microscope only) and dinoflagellate cyst mapping can help to refine monitoring strategies but can never be used to make management decisions. Locally designed models which incorporate oceanographic and climatic conditions (e.g., ENSO events) can provide useful indices to predict red tides. For example, two weeks of calm weather and sunshie usually precede *Pyrodinium* blooms in Sabah. We need to learn from every red tide event to improve monitoring strategies, change aquaculture methods (e.g., intertidal oysters become less toxic than longline mussels) or even change dietary habits (e.g., Japan has a very limited mussel consumption).

Communications

Effective communication networks exist between individual scientists but are not in place at government levels. Multilateral communication networks need to be established for dissemination of information. ICLARM, which is an international, nongovernmental organization, was suggested as a suitable base, where people can fax information to be disseminated in a red tide newsletter (3 times per year). Responsibilities of Public Health, Environment and Fisheries Departments need to be clarified. In Canada and Tasmania, Fisheries officers carry out the monitoring but Public Health ministers issue red tide warnings.

Discussions and Recommendations on Management Issues

Chairperson: Chua Thia-Eng

The Working Group on *Pyrodinium* red tide management focused on the necessary responses that should be taken on the part of the concerned government agencies in times of red tide outbreaks. The Working Group noted the *ad hoc* nature of most governments' responses to red tide outbreaks and stressed the importance of developing a national policy and contingency plans to minimize the impact of PSP on human life.

Policy

The group noted that there were over 60 fatalities and more than 1,000 cases of people taken ill due to consumption of contaminated shellfish since 1976. Further illnesses could be prevented if appropriate actions are taken by the government to warn the public and develop appropriate treatment for PSP victims. Recognition by the government of the severity of red tide outbreaks to public health and the economy is very essential. The Government of Brunei Darussalam, for example, has established the policy of containing red tide without compromising human lives.

The Group suggests the following policy be adopted by governments:

"To prevent loss of human lives and minimize economic losses arising from red tides."

Approaches

The Group recognized the fact that very little can be done to economically and effectively control the outbreak of red tides.

However, it was agreed that appropriate measures can be taken to manage the consumers from taking contaminated seafoods thereby minimizing the adverse effects of red tides.

Management Measures

Management measures in the form of action plans should be developed by government implementing agencies in order to mitigate the adverse effects of red tide outbreaks.

The following measures are considered essential:

1. Monitoring and Surveillance

Efforts should be made to undertake a systematic monitoring and surveillance program to more accurately detect or predict the occurrence of red tides. This could be done through periodic sampling of plankton for the presence and abundance of red tide organisms or their cysts in the water column and the sea bed. The level of toxicity in planktivorous fish and mussels should also be monitored. Usually the fisheries agency takes the main role in monitoring and surveillance. The need to establish adequate sampling stations and standardize sampling and analysis procedures was stressed and the methodologies are provided in the training manual (Part 2) of this document.

2. Regulatory Measures

Measures to regulate harvesting and marketing of shellfish during red tide outbreaks are very essential in order to reduce the chance of contact with consumers. Regulatory measures include:

- imposing/lifting bans on harvesting and marketing of shellfish and other products contaminated by red tide organisms;
- destruction of contaminated aquatic products;
- control of sales of contaminated aquatic products at retail or wholesale markets.

3. Public Awareness and Education

Members of the public including policymakers and administrators need to be made aware of the adverse impacts of red tides on human lives and economic loss arising from red tide outbreaks. Concerted efforts should be made by concerned government agencies to (a) increase general knowledge and concerns of the public on the impacts of red tides through adequate use of media and other mass communication channels; and (b) solicit public support, cooperation and confidence in management measures undertaken by the relevant authority.

4. National Coordinating Committee and Response Teams

The above measures and actions need to be adequately and effectively coordinated by a central body and implemented by one or more response teams depending on the estimated area or scope of possible impacts. A single national coordinating committee is very necessary especially in big nations where the impact area could be large. Countries like the archipelagic Philippines will certainly need more than one response team. The national committee will have the overall responsibility to mitigate the adverse effects of red tides. The committee should be made up of the various concerned agencies (such as Fisheries, Environment and Natural Resources, Public Health, Port and Harbor, and Marine Department, etc.), non-government organizations (NGOs) and representatives from affected coastal communities. The response teams shall undertake field monitoring, surveillance and enforcement.

5. National Capability to Respond to Red Tide Outbreaks

National capability in terms of manpower and facilities should be considerably improved or strengthened in order to be efficient and effective in executing national red tide contingency action plans. There should be sufficient technical manpower for monitoring the occurrence of red tides. The availability of monitoring and surveillance facilities such as vessels, sampling and analyzing equipment to accurately detect red tide outbreaks on time will ensure appropriate mitigating measures. Sufficient medical backup for treatment of PSP is also essential.

6. Public Warning

One most effective measure in response to red tide outbreaks is preventing the public from consuming contaminated seafood. Public warning strategies are essential and should be incorporated into red tide response contingency plans. These plans should include an effective public warning mechanism to inform the public of PSP and what necessary actions are to be taken. Appropriate timing in making public announcements is necessary and should be guided by the level of PSP. The content of the public announcements should be specific in terms of prevention and regulatory measures approved by the authority concerned. Public announcements should be coursed through the most effective media (TV/radio/newspaper) or village or community contacts. While public warning or awareness is a crucial step in the implementation of the contingency plan, overpublicity could be counter-productive and may create public panic and adverse marketing effects of suspect seafood. The public should have confidence in their government's actions.

Contingency Action Plans

Establishing of an administration and implementing structure is a first step in launching a contingency plan. The national coordinating committee and response team(s) must be established in order to execute the functions of coordination with concerned implementing agencies in performing monitoring and surveillance duties.

Information on the occurrence of red tide organisms or indication of PSP can be obtained from fishermen, fishermen's associations coastguard, marine police or through the red tide monitoring unit(s). All information received should first be confirmed before the other actions of the contingency plans are put to motion. An aerial survey is helpful to determine the spread of red tides. The monitoring unit must be placed on full alert to continue monitoring for red tide indications even if the report is not confirmed. When red tide occurrence is confirmed, monitoring activities would have to be intensified and medical and health services readied for possible PSP victims. The coastguard, marine police and fishermen's associations should be informed and requested to continue reporting to the response team for possible occurrence of red tides in other water bodies.

Market surveillance by the fisheries department and municipal, medical and public health authorities is another crucial step in the plan to minimize the consumption of contaminated seafood. Whenever necessary, regulatory control measures or prohibition on harvesting and marketing of contaminated aquatic products should be implemented. The public should then be warned through the appropriate communication channels.

Transnational monitoring is essential amongst neighboring nations affected by red tide outbreaks.

A generalized contingency plan based on the above points is illustrated in Fig. 1.

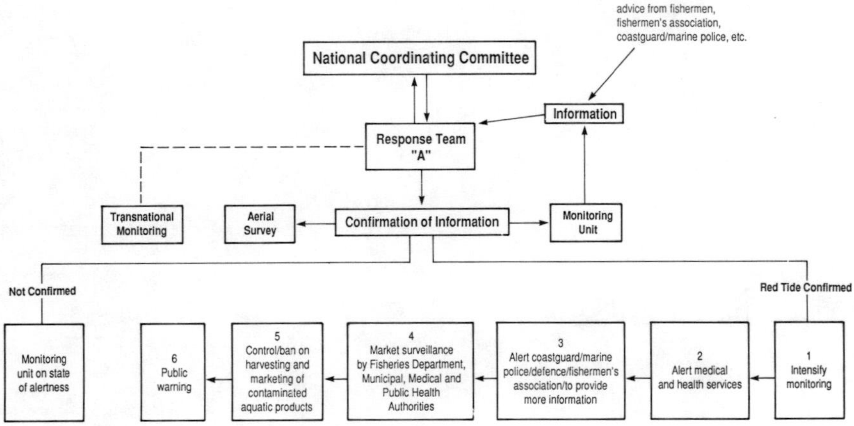

Fig. 1. A schematic illustration of actions of red tide contingency plans.

Epidemics of Paralytic Shellfish Poisoning in the Philippines, 1988-1989

NIÑO ISMAEL S. PASTOR
IVY GOPEZ
MA. CONSORCIA QUIZON
NICOLAS BAUTISTA
MARK WHITE
MANUEL DAYRIT
Field Epidemiology Training Program
Department of Health
Philippines

PASTOR, N.I.S., I. GOPEZ, M.C. QUIZON, N. BAUTISTA, M. WHITE and M. DAYRIT. 1989. Epidemics of paralytic shellfish poisoning in the Philippines, 1988-1989, p. 165-171. *In* G.M. Hallegraeff and J.L. Maclean (eds.) Biology, epidemiology and management of *Pyrodinium* red tides. ICLARM Conference Proceedings 21, 286 p. Fisheries Department, Ministry of Development, Brunei Darussalam, and International Center for Living Aquatic Resources Management, Manila, Philippines.

Abstract

Health-related aspects of six epidemics of paralytic shellfish poisoning (PSP) in the Philippines between July 1988 and February 1989 are described. Results of epidemiological investigations showed mussels and scallops to be associated with PSP; the incubation period was longer than experienced elsewhere; and PSP cases occurred well below the recognized limit of toxin for consumption of 80 μg/100 g of meat. Recommendations to study these problems and for a national shellfish sanitation program are made.

Introduction

Since 1972, there have been increasing reports of red tides in the Indo-Pacific basin (Maclean and White 1985). This biological event is correlated with human illness called paralytic shellfish poisoning (PSP) and can have severe impacts on local fisheries.

The Philippines sustained seven documented epidemics of PSP between 1983 and 1987. A total of 978 patients with a case-fatality ratio of 2 per 100 cases was reported (Arteche 1984; Estudillo and Gonzales 1984; Gacutan et al. 1985; Varona 1987). The epidemics were caused by *Pyrodinium bahamense* var. *compressum*.

This paper presents the results of the epidemiological investigations of subsequent PSP epidemics between 1988 and 1989 in different parts of the Philippine archipelago.

Methodology

The Department of Health investigated 437 reports of PSP suspects. To confirm these reports, investigators interviewed the suspects themselves. A standard questionnaire was applied (Appendix I). These were also used among neighborhood or household controls matched for age range (± 5 years) and time of meal of the suspects. A case-control study identified risk factors using McNemar's paired test at the 95% confidence interval.

A case definition was developed. It was adopted from the definition by the Centers for Disease Control during a *Pyrodinium*-caused PSP in Champerico, Guatemala, in 1984 (Dean et al. 1988). A patient was defined as a previously healthy person who suddenly developed at least two of any of the following motor defects: inability to walk properly, dysphagia, dysphonia, dyspnea, paralysis, weakness. In addition, the patient should also have two of any of the following sensory defects: numbness, paresthesia, feeling hot (febrile), pruritus, dysthesia, light-headedness, and short-tongue sensation.

Results and Discussion

There were six outbreaks (Table 1) beginning 9 July 1988 and ending 14 February 1989, affecting seven provinces nationwide (Fig. 1).

A total of 224 patients fitted the case definition. This corresponds to moderate and severe PSP. Mild cases could not be differentiated from hysteria and other marine biointoxications. For example, on 25 August 1988, epidemic surveillance reported 19 PSP suspects which turned out to be ciguatera cases from Basilan Island, about 1,000 kilometers south of Manila (Guerrero 1989).

Table 1. Profile of six Philippine PSP outbreaks (1988-1989).

Province	Cases	Death	Epidemics start	Dates end	Toxin*	Transvector	Mean onset (hours)
Bataan	44	1	8/19/88	9/30/88	1,005	Cultured mussels	4
Manila	14	2	9/06/88	9/30/88	80 to 90	Cultured mussels	4
Cavite	8	1	9/09/88	9/10/88	40 to 100	Cultured mussels	15
W. Samar	22	0	7/09/88	9/05/88	164	Wild mussels	4
Negros	109	4	12/14/88	12/19/88	47	Cultured mussels	9
Capiz	3	1	12/14/88	12/14/88	(?)	(?)	(?)
Cebu	24	5	2/01/89	2/14/89	2,000	Cultured scallops	6
Total	224	14					

*Micrograms of toxin per 100 g of sample obtained initially
(?)Mussels causing illness came from floating bamboo poles

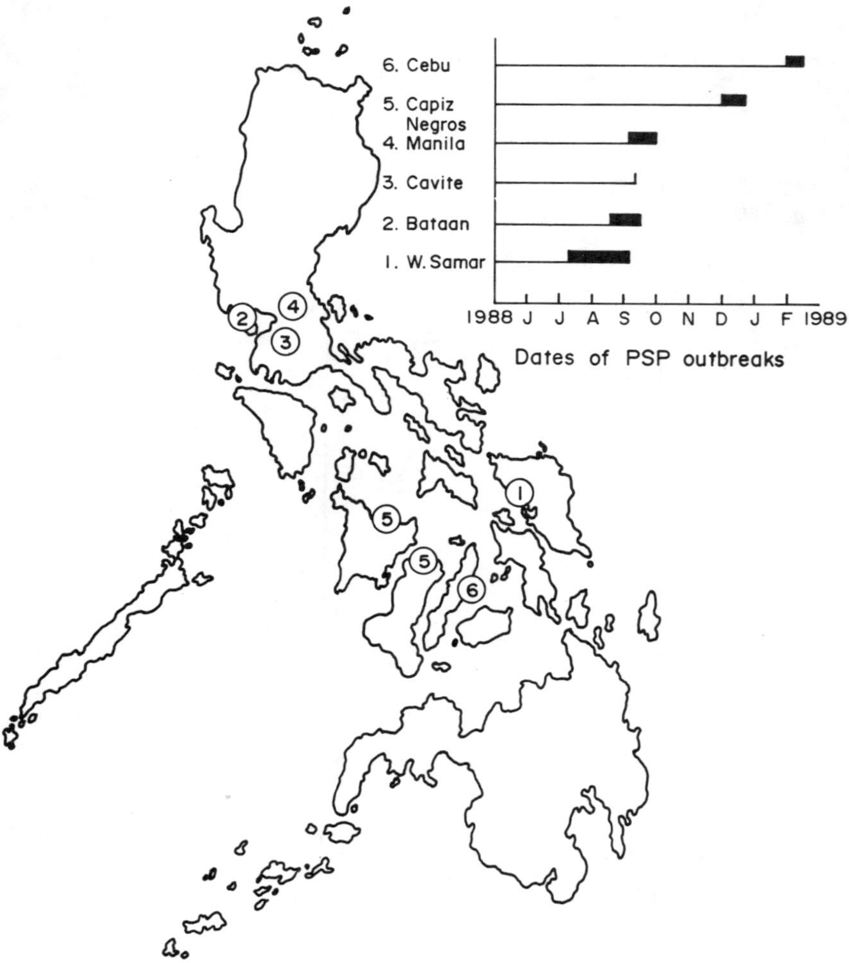

Fig. 1. Sites of PSP, Philippines, 1988-1989.

The ages of PSP cases ranged from one to 78 years with a mean of 29 years. The age-specific attack rates were highest between the 30 and 40 years group reflecting greater exposure in this age group. Eighty-six per cent of deaths were of children aged 13 years and below. The case-fatality ratio was 6 per 100 patients.

The incubation period, the time from ingestion of shellfish until the onset of any symptom, was from almost immediately to as long as thirty-four hours. The median was 5.6 hours. This was longer than the expected median onset for PSP which is 30 to 45 minutes with a usual range of 30 minutes to three hours (Hughes 1979).

The symptoms involved the gastrointestinal, neural and the respiratory systems (Table 2). The illness started with vomiting followed by numbness with or without paresthesia of the circumoral area. Patients reported that the numbness descended to the upper then lower extremities. This led to a light-headed or floating sensation. Later, inability to walk properly was experiénced followed by dyspnea, dysphagia, and dysphonia in this order. Deaths were due to respiratory failure.

Table 2. Symptoms of PSP cases Philippines, 1988-1989 (N = 224).

Symptoms	Per cent
Gastrointestinal	
Vomiting	55
Abdominal pain	34
Water diarrhea	15
Nausea	8
Hypersalivation	4
Motor Abnormalities	
Inability to walk	71
Paresis of extremities	45
Dyspnea	44
Dysphagia	34
Diplopia	20
Paralysis	17
Dysphonia	7
Sensory Abnormalities	
Numbness	90
General body malaise	80
Dizziness	73
Light headed sensation	54
Headache	48
Paresthesia	40
Felt hot	16
Dysthesia	12
Short tongue sensation	11
Pruritus	4

Case-control studies revealed that green mussels *(Perna viridis)* and Asian moon scallops *(Amusium pleuronectes)* posed the greatest risk for PSP (Table 3). Shrimps, fish with gills and guts (a common practice among Filipinos), and sergestid shrimps did not pose a risk.

The results of the animal inoculation tests on initial samples from farms supplying epidemic sites are shown in Table 1. The

Table 3. PSP transvectors.

Marine products	Odds ratio	95% confidence interval
Assorted fish	0.1	0.1 to 0.2
Shrimps	0.5	0.2 to 1.2
Green mussels	16.3	9.8 to 27.3
Asian moon scallops	8.3	1.9 to 40.0

investigators were unable to recover the offending food items or the gastric contents, serum or urine from the patients.

To prevent epidemics, various forms of information dissemination strategies were tried. One form was a lecture to a target group consisting of private citizens. Another was a lecture to physicians, nurses, sanitarians and local public officials regarding the features of the illness.

The more successful health information strategies were the red tide "updates: quarter-page advertisements in national as well as local newspapers issued by member of the red tide monitoring team consisting of biologists, food chemists, physicians, lawyers, press relations officers and managers from the Department of Health and the Bureau of Fisheries and Aquatic Resources.

Recommendations

We propose a national PSP control and red tide management program. The activities of this program should be toxicological monitoring of shellfish and made available to distant island provinces within the country. This should be aimed at upholding the health and safety of the public while protecting the reputation of the shellfish industry.

Studies should also be aimed at describing the life history of *Pyrodinium bahamense* var. *compressum* including ecological factors.

In addition, the public should also be primed with information about red tide during red-tide-free seasons. This will prepare the general public for what to expect when it happens.

Acknowledgements

Our thanks go to Mr. Jay Maclean (ICLARM), Mr. Cielito Gonzalez (BFAR), Dr. Nelia Maramba (University of the Philippines), and all personnel of the Bureau of Food and Drugs of the Department of Health for aiding us during the height of the epidemics.

References

Arteche, L.O. 1984. Epidemiological research on red tide occurrence in June to October 1983. Integrated Provincial Health Office, Catbalogan, Samar. Unpublished report. 16 p.

Dean, J.A., A.G. Dean, A. Burton and R. Dicker. 1988. Epiinfo Version 3.0: computer programs for epidemiology. Centers for Disease Control, US Department of Health and Human Services, Atlanta, Georgia.

Estudillo, R.A. and C.L. Gonzales. 1984. Red tides and paralytic shellfish poisoning in the Philippines, p. 52-79. *In* A.W. White, M. Anraku and K.K. Hooi (eds.) Toxic red tides and shellfish toxicity in Southeast Asia. Southeast Asian Fisheries Development Center, Singapore, and International Development Research Centre, Ottawa, Canada.

Gacutan, R.O., M.Y. Tabbu, E.J. Anjero and F. Icatlo. 1985. Paralytic shellfish poisoning due to *Pyrodinium bahamense* var. *compressa* in Mati, Davao Oriental, Philippines. Marine Biology 87: 223-227.

Guerrero, E.T. 1989. Fish poisonings in Basilan. Field Epidemiology Training Program, Department of Health. Philippines. Unpublished manuscript. 14 p.

Hughes, J.M. 1979. Epidemiology of shellfish poisoning in the United States, 1971-1977, p. 23-28. *In* D.L. Taylor and H.H. Seliger (eds.) Toxic dinoflagellate blooms. Elsevier - North Holland, New York.

Maclean, J.L. and A.W. White. 1985. Toxic dinoflagellate blooms in Asia, a growing concern, p. 517-520. *In* D.M. Anderson, A.W. White and D.G. Baden (eds.) Toxic dinoflagellates. Elsevier Science Publishing, New York.

Varona, W. 1987. Report to the Undersecretary of Public Health, Department of Health. Tacloban: Regional Health Office. No. 8, 22 September 1987. Unpublished report.

Discussion on Medical and Epidemiologic Issues

Chairperson : N.I. Pastor
Rapporteur : R.A. Corrales

The session opened with a list of problems that included the incubation period for PSP, the reporting system, the intervention strategies, identification of the food vehicle, diagnosis of paralytic shellfish poisoning, and correct sampling techniques.

Incubation Period

The median incubation period of 5.6 hours in the Philippine PSP epidemics of 1988 to 1989 is much longer than that reported elsewhere.

The incubation period was the time that lapsed from the offending meal to the time when patients developed symptoms of PSP. The group received the impression that this was the time from the meal until the time when the patients fit the case definition developed in the Philippines. A case was defined as a person who was previously healthy and who developed two motor and two sensory abnormalities. It was felt that this was a very strict case definition, corresponding to Prakash et al. (1971) moderate to severe PSP. This difference may have accounted for the longer incubation period. However, only sustained epidemiologic investigations will resolve the problem.

Reporting System

The problem of underestimating and/or overestimating the actual number of PSP victims was discussed. Hospital admissions were used as a late indicator of a PSP outbreak in the Philippines. The number of admissions did not reflect the true incidence of PSP since many patients preferred to stay home. The investigators conducted a survey once community clusters from hospital admissions were identified. In the community, many residents attributed all kinds of ailments to the red tide. This may have led to an exaggerated number of PSP victims. The case definition was developed to avoid false positives and negatives.

The group endorsed a community survey using a standardized questionnaire and case definition for *Pyrodinium*-caused PSP. Protocols for investigating PSP epidemics should also be developed and made available to the region.

The group addressed the problem of increasing reports of PSP in the region. It may be a result of increased awareness, rapid information dissemination channels, and better detection technology. In new epidemics without baseline data, a retrospective community study would be very helpful. This would be the epidemiological counterpart of paleoecological studies on core samples. Certainly in the Philippines, more provinces are reporting PSP than in previous years.

Intervention Strategies

Three levels were discussed: preventing distribution of contaminated shellfish; preventing consumers from ingesting contaminated shellfish; and preventing deaths among persons who had ingested contaminated shellfish. The proposed strategies to deal with these issues were, respectively: quarantine and closures of contaminated farms; informing the public; and emergency remedies and a therapeutic regimen for moderate to severe PSP.

The group recommended that to avoid confusing the public, a single agency should be authorized to impose and lift quarantines. After a long discussion on the acceptability of the AOAC standard, the group endorsed the toxic level of 80 micrograms of saxitoxin standard per 100 grams of shellfish, pending re-evaluation of *Pyrodinium* toxicity. It was further proposed that standards for other types of red tide toxins should be developed and made available to other laboratories.

The chairperson presented the various emergency measures and therapeutic regimens tried in the Philippines and elsewhere. Among the household remedies encountered in the Philippines, coconut milk (the whitish extract obtained by squeezing shredded coconut meat) with brown sugar and coconut oil were the most popular. The extract is a popular household remedy among coastal residents in the region for all kinds of ailments. It has been institutionalized in Philippine PSP-endemic provinces as the most effective household remedy. Its mechanism of action is not clear.

The Philippine Department of Health had also tested several substances to minimize the effects on mice, chicks, rabbits and puppies given diluted standard saxitoxin solutions per orem (Roxas and Barreyro 1984). Consistent results were obtained from activated charcoal, powdered coconut shell charcoal, and coconut milk given along with saxitoxin solutions per orem to animals which eventually recovered. Various herbal preparations such as *Moringa oleiferia* Lam, and *Sesbania grandiflora* (Linn.) Pers. yielded inconsistent results. Clearly, further research on the various first aid materials is needed to ascertain their specific actions.

The physicians of the Philippine Department of Health have also tried several drugs in their patients. Among these drugs were steroids, sodium bicarbonate, sodium sulfate, atropine sulfate, loop diuretics, physostigmine, and neostigmine. Health officers in Sabah during the 1976 PSP outbreak employed diuretics, anticurare drugs, oximes, noradrenalin, ephidrine and amphetamine (Roy 1977).

The therapeutic regimen developed at the Philippine General Hospital (Samar et al. 1988) was based on the knowledge that red tide toxins are stable in acid and soluble in water. Loop diuretics and alkaline lavage, in addition to intravenous administration of sodium bicarbonate, and physostigmine with atropine sulfate, may be promising. The latter drugs are used as adjuncts during assisted ventilation. Loop diuretics may hasten the elimination of the water soluble toxin. An alkaline lavage may "neutralize" the acid-stable toxin. Otherwise the red tide toxin at present has no known antidote. Treatment is usually supportive including gastric lavage, gastric evacuation, and catharsis to eliminate the unabsorbed toxin. Prognosis improves with artificial respiration. The treatment protocol used at the Philippine General Hospital is shown in Fig. 1.

Identification of food vehicles

Marine fisheries products can contain high levels of PSP yet do not qualify as a transvector. This happens when a product is not commonly eaten by humans.

The criterion for listing a marine fisheries product as a transvector for PSP was its implication by epidemiological and laboratory analysis as the cause of human illness.

Epidemiological analysis involves calculating the risk of developing symptoms following a meal consisting of several food items. The purpose is to determine which among the food items eaten during the suspect meal caused the symptoms. For example, a typical Filipino meal consists of rice, water and "viands"-meat and/or vegetable dishes. For example, coastal residents may eat in a single meal scallops, mussels and uneviscerated fish all of which can be PSP transvectors. Risk analysis will determine which food items actually caused the illness.

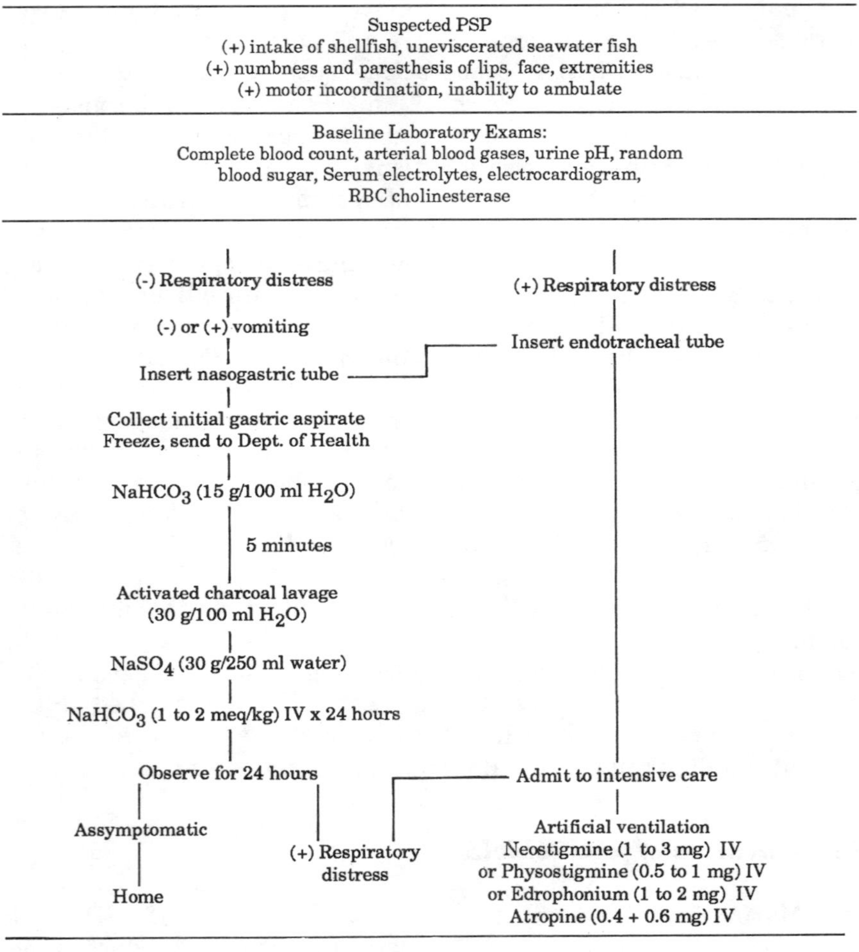

Fig. 1. PSP treatment algorithm at the Philippine General Hospital, courtesy of Drs. Samar, Dioquino and Maramba, Philippine General Hospital.

Mouse bioassay of the offending food item(s) determines whether a marine fisheries product is a transvector. Unfortunately most patients discard the poisonous food item(s). Investigators in the Philippines fall back on the patient's dietary history. When a possible transvector is identified from the history, similar samples from those places where patients and vendors obtained the offending product(s) are sent to the laboratory. Tracing the flow of food from a patient back to a farm or sea bed is a tedious process that can last several days of investigative work, especially in areas affected for the first time.

Experience in the Philippines showed that marine fisheries products should be included in the list of transvectors when results of epidemiological and mouse bioassay analyses were consistent.

Diagnosis of PSP

PSP is diagnosed on clinical grounds along with a positive history of intake of an established transvector.

The symptoms of PSP refer to a combination of neurological respiratory and gastrointestinal abnormalities. Three stages had been identified by Prakash: mild, severe, and extreme. Mild cases may be difficult to differentiate clinically from conversion reactions and ciguatera. Severe cases were initially diagnosed as cerebro-vascular accidents. The dietary history provides very helpful clues. But in areas of first occurrence, this becomes difficult. Patients do not know that their symptoms were caused by their meal. The Philippine investigators then developed the diagnostic criteria mentioned above.

Sampling Techniques

Saxitoxin should be demonstrated from the offending meal, then from the gastric contents and the serum or urine of patients who ate this meal. PSP and its behavior inside the human body can then be studied.

Unfortunately, during most outbreaks none of these samples were obtained. In addition, no study had addressed the pharmacodynamics and pharmacokinetics of saxitoxin. It is not known if saxitoxin can be recovered from serum or urine. A Singapore study demonstrated PSP in the gastric content of one out of two patients who ate green mussels.

Future investigations should attempt to demonstrate saxitoxin from the offending meal, gastric content, serum or urine of patients. Experiments using animal models may also reveal how mammals absorb, transform and eliminate PSP. Such experiments are planned at the College of Medicine of the University of the Philippines.

References

Prakash, A., J.C. Medcof and A.D. Tennant. 1971. Paralytic shellfish poisoning in eastern Canada. Bull. Fish. Res. Board Can. 177.

Roxas, M.G. and D.A. Barreyro. 1984. Final report of the research project: A study on possible neutralizers of the toxin ingested by the *Tahong* (mussels) and other shells from Maqueda Bay, Western Samar. Department of Health, Tacloban, Philippines. 80 p. (mimeo).

Roy, R.N. 1977. Red tide and outbreak of paralytic shellfish poisoning in Sabah. Med. J. Malays. 31(3): 247-251.

Samar, O.M., C. Dioquino and N. Maramba. 1988. Neurotoxic manifestations of "red tide" poisoning. Paper presented at the Neurology Staff Conference, Department of Neurology, Philippine General Hospital, Manila, Philippines. September 20, 1988. 5 p. (mimeo).

Economic Aspects of *Pyrodinium* Red Tides in the Western Pacific

J.L. MACLEAN
International Center for Living Aquatic Resources Management
MC P.O. Box 1501
Makati, Metro Manila
Philippines

MACLEAN, J.L. 1989. Economic aspects of *Pyrodinium* red tides in the western Pacific, p. 179-185. *In* G.M. Hallegraeff and J.L. Maclean (eds.) Biology, epidemiology and management of *Pyrodinium* red tides. ICLARM Conference Proceedings 21, 286 p. Fisheries Department, Ministry of Development, Brunei Darussalam, and International Center for Living Aquatic Resources Management, Manila, Philippines.

Abstract

For economics purposes, red tide areas can be grouped into chronic and acute situations and areas of permanent closure, respectively. Different kinds and levels of costs are involved in each. *Pyrodinium* red tides have largely been treated as acute situations, with occasional large financial losses. The economic implications of dealing with them as chronic situations are discussed and it is suggested that permanent closure of some areas may be more cost effective.

Introduction

Of all the aspects of red tides, economic consequences even more so than health aspects are probably the most important to politicians and individual fishermen. Yet, they are the least studied. By way of background, the limited literature on this subject exhibits different kinds of economic impact of red tides. These are costs/losses associated with chronic situations, acute situations and permanent closures, respectively.

Losses Due to Chronic Red Tides

Losses of fish due to red tides in the Seto Inland Sea, Japan, began in 1970. In 1972, *Chattonella* red tides killed 14 million caged yellowtail fish there, corresponding to a loss of 71 billion yen (Okaichi 1985). A compilation by prefecture shows that over 1,000 businesses were affected, in dollar terms a loss of nearly $24 million. Losses in subsequent years to 1976 averaged about $1 million. However, to put this in perspective, 80-90% of aquaculture losses in the Inland Sea of Japan at the time were due to oil pollution (Sakiyama 1979).

Okaichi (1989) in a more recent review shows that annual losses from red tides have been creeping up again - to 2.1 billion yen in 1987, although red tide incidents have decreased since 1976.

In Hong Kong, 11 fish kills due to red tides from 1980 to 1984 caused a total loss of HK$4.2 million (Wong and Wu 1987), while molluscan shellfish kills from brown tides in New York during much the same period resulted in economic losses of about $2 million/year (Kahn and Rockel 1988).

In passing it is noteworthy that ciguatera, which is also caused by dinoflagellates, is responsible for about 2,300 cases of fish poisoning each year in north America, costing up to $20 million in time off work and hospitalization (Gervais and Maclean 1985).

Also associated with chronic situations are the costs of monitoring, primarily by regular toxin testing in selected sites. Attention was paid to this issue during the Second International Conference on Toxic Dinoflagellate Blooms (Taylor and Seliger 1979). Some figures were provided during a workshop on impact of blooms on the north American shellfish industry. In Maine, a regular sampling program then cost $20-40,000/year to service an $8 million shellfishery. The "in-state" added value was said to be actually $40-80 million (in the late 1970s). Even so, there was a loss in excess of $7 million in 1980 in Maine due to PSP (Shumway et al. 1988). In New Hampshire, toxin monitoring cost about $10,000/year at the time, for a recreational shellfishery generating $50-60,000/year in license fees. Market value harvested was $180-200,000/year. In Canada's Maritime Provinces, it was costing over $50,000/year to monitor shellfisheries worth about $1 million/year in 1973 (Taylor and Seliger 1979).

Losses in Acute Situations

A 1980 PSP outbreak in the San Francisco, California, area has interesting parallels with a 1988 Manila Bay, Philippines, outbreak (see below). Toxin levels in oysters exhibited levels above 80 µg/100 g for up to 22 days over about 100 km of coastline, mostly north of San Francisco. The problem began in late July 1980 but a ban on harvesting in this area was lifted by late August. Strong media coverage and misleading information resulted in nearly 100% closure of markets in California and 25% of those even in Oregon and Washington. Markets were still depressed by Christmas 1980. Losses by growers amounted to $630,000 (Conte 1984).

Losses to the tourist industry in occasional severe Florida red tides have been substantial: $18 million in 1971; $15 million in 1973-74 (with additional costs to the real estate industry); and a recurrence would now cause losses of $1.25 million per day (Taft and Martin 1986).

Towing 120 farms out of threatened areas during the severe red tides in Norway in 1988 *saved* around $200 million worth of fish (Anon. 1988).

Permanent Closure

Permanent closure of much of British Columbia's Pacific coastline to shellfishing due to chronically high toxin levels was said to have caused annual potential losses of $2 million, not to mention the loss of use of the resources themselves. The ban on sale of suspected ciguatoxic fish species in Florida, the Caribbean and Hawaii was estimated at costing $10 million/year for lost business (Taylor and Seliger 1979).

Impact of *Pyrodinium*

We tend to think of the various *Pyrodinium* incidents as acute rather than chronic situations because of their irregularity. Thus, their impact is measured in isolation and the few estimates of costs/losses have not taken into account the effects of shifts in consumer preference, product substitution, etc. (Kahn and Rockel 1988). With these caveats, following are somewhat subjective analyses of the costs of *Pyrodinium* events in the western Pacific.

Brunei Darussalam

The first outbreak in 1976 caused losses of about US$100,000 in condemned seafood. Government welfare was also provided over a two-month period (Beales 1976). A second bloom in 1980 possibly had similar economic consequences. At present, mussel culture continues at a small scale. Its potential is directly affected by chronically high toxin levels in most shellfish since 1980. In fact, high toxin levels in mussels during 1988 resulted in losses of about $60,000 to the sole mussel farm (Matdanan and De Silva, this vol.).

Shrimp culture is planned - up to 5,000 ha of ponds (Chua et al. 1987). However, red tide "scares" may occasionally and/or chronically affect exports. (So may *Chattonella* outbreaks, which have caused heavy mortality in Malaysian shrimp ponds (Khoo 1985)). A proposed action plan (Matdanan et al. 1988; De Silva et al., this vol.), when implemented, can be considered a chronic situation expense.

Papua New Guinea

There was no commercial mollusc shellfishery in red tide areas of PNG. Fatalities as a result of the 1972 *Pyrodinium* season in Port Moresby did not prevent villagers gleaning shellfish in following *Pyrodinium* seasons, despite publicity. Major costs of the seasonal blooms during 1972-1974 were for research and monitoring of about US$30,000/year. Thereafter, there was no further activity or expense.

Philippines

The first blooms in the Samar Sea in 1983, were said to have cost the mussel industry there $5 million "not to mention loss by the fishing industry as a result of the scare" (Gonzales et al. 1989), Possibly the same level of loss occurred in the subsequent outbreaks there in 1987 and 1988.

In August and September 1988, the first outbreak in Manila Bay occurred. Thanks to the media, the whole seafood industry nearly ground to a halt, while mussel growers even tried to implicate freshwater products in an effort to offset the swing by consumers to tilapias and other freshwater organisms! All fish markets in Manila were depressed for over three months, similar to the case in San Francisco in 1980. Manila's seafood market handles 35% of the nations landings. Thus, the losses were large, up to $300,000/day at the height of the scare. Japan and Singapore banned shrimp imports from the Philippines for an unknown period (although they were "clean"), which would have meant losses of $500,000/day if the produce was not subsequently sold. Losses by mussel growers for a three-month period were more modest, about $950,000 in all (Maclean 1989).

Sabah

The 1976 and 1980 red tides that affected Brunei Darussalam stretched north for 200 km through Sabah. Since a number of people died and the red tides were far more extensive than in Brunei Darussalam, economic losses must have been proportionally larger.

Another resource was badly affected also, the coral reefs. Elizabeth Wood (unpublished data) monitored the devastation of inshore reefs. Six months after the blooms, 90% of the coral substrate was dead. Invertebrates took many months to recolonize the area. The loss in terms of fisheries produce is very difficult to estimate.

Although there have been no visible blooms in Sabah since 1980, there have been several deaths from PSP each year, which would eliminate any confidence in building a molluscan shellfishery; in fact, oysters in an experimental plot at Kuala Penyu were always toxic (Ting Thian Ming, pers. comm.; Ting and Wong, this vol.).

Sabah has plans to expand its shrimp culture industry and officials have played down the red tide issue to avoid a scare which would depress shrimp exports. A figure for exports of $28 million/year was given (Lee 1988).

Discussion

The few observations above are inadequate to indicate whether monitoring is cost effective. Large losses can still occur in acute situations in a monitored shellfishery. However, the biggest such losses seem to be in tourism, which is not an issue with *Pyrodinium,* and the result of misinformation, as in San Francisco and the Philippines.

The economic effects of *Pyrodinium* blooms have been moderate in comparison with those caused by other species in other regions, with the exception of that in Manila Bay, Philippines, where the major impact was on the "clean" marine fisheries.

In the *Pyrodinium* zone, chronically high toxin levels in shellfish on the west coast of Borneo are indicative of chronic situations that require either permanent closure or perpetual monitoring. The irregular outbreaks in the Philippines also suggest that there will be recurrences. Monitoring in a number of locations is indicated.

To plan on the basis that *Pyrodinium* red tides are becoming chronic situations that require monitoring has several economic implications:

1. A team has to develop an action plan.
2. Implementation of the plan has to be considered a "permanent" expense.

3. A continuing public awareness campaign to maintain alertness is needed.
4. Occasional bans on harvest and enforcement of such bans must be carried out.
5. A mechanism to end bans is needed; coordination between government agencies is critical.

Some indirect consequences may be:

6. Loss of public confidence in seafoods from the monitored area.
7. General loss of confidence, for example, by overseas importers, in an affected country's seafoods.

On the other hand, monitoring may in the long term *promote* confidence in the products. To a large extent this will depend on whether bans that may be imposed are enforceable and/or respected. Therefore, the prognosis is not good in the Philippines at least.

If a ban is not going to be respected, it only adds to the costs of coping with the occasional "acute" situations. There may be loss of life as if the monitoring and warnings were not carried out. I am afraid that if bans are not respected, monitoring is a fruitless, academic exercise.

If monitoring is not useful, then each period of high toxicity in seafoods remains a separate acute economic (and health) crisis. The only alternative is a permanent closure of identified areas. In Borneo this would in effect end the bivalve shellfish industry. In the Philippines, it would require dismantling the mussel industry in affected areas and moving it to safer waters. It seems to me that such rather drastic action would be the most cost effective, but more data are needed to resolve the matter.

References

Anon. 1988. Over 100 farms towed to avoid algal threat. Can. Aquacult. 4(5): 43.

Beales, R.W. 1976. A red tide in Brunei's Coastal waters. Brunei Mus. J. 3: 167-182.

Conte, F.S. 1984. Economic impact of paralytic shellfish poison on the oyster industry in the Pacific United States. Aquaculture 39: 331-343.

Chua, T.E., C. M. Chou and M.M. Sadorra, editors. 1987. The coastal environmental profile of Brunei Darussalam: resource assessment and management issues. ICLARM Tech. Rep. 18, 193 p.

Gervais, A.J. and J.L. Maclean. 1985. Management of fisheries and public health problems associated with toxic dinoflagellates, p. 530-533. *In* D.M. Anderson, A.W. White and D.G. Baden (eds.) Toxic dinoflagellates. Elsevier, New York.

Gonzales, C.L., J.A. Ordoñez and A.M. Maala. 1989. Red tide: the Philippine experience, p. 43-46. *In* T. Okaichi, D.M. Anderson and T. Nemoto (eds.) Red tides, biology, environmental science and toxicology. Elsevier, New York.

Kahn, J.R. and M. Rockel. 1988. Measuring the economic effects of brown tides. J. Shellfish Res. 7: 667-682.

Khoo, E.W. 1985. Occurrences of "red tide" along Johore straits, Malaysia, resulted in heavy mortality of shrimp. World Maricult. Soc. Newsl. 16(1): 4.

Lee, S.J. 1988. "Red tide" menace endangers fisheries. Jakarta Post 23 April 1988.

Maclean, J.L. 1989. Indo-Pacific red tide occurrences, 1985-1988. Mar. Pollut. Bull. 20: 304-310.

Matdanan, Haji Jaafar, De Silva, M.W.R.M., Sharifuddin. P.H.Y. and Siti Amin Mahali. 1988. A red tide action plan for Brunei Darussalam. Paper presented at the Technical Workshop on Integrated Tropical Area Management, Singapore, 28-31 October 1988. National University of Singapore, Science Council of Singapore, and International Center for Living Aquatic Resources Management, Manila, Philippines.

Okaichi, T. 1985. Fish kills due to the red tides of *Chattonella.* Bull. Mar. Sci. 37: 772 (Abstract).

Okaichi, T. 1989. Red tide problems in the Seto Inland Sea, Japan, p. 137-142. *In* T. Okaichi, D.M. Anderson and T. Nemoto (eds.) Red tides, biology, environmental science and toxicology. Elsevier, New York.

Sakiyama, T. 1979. The impact of expanding environmental pollution on Japanese fisheries in the coastal zone, p. 345-355. *In* A.R. Librero and W.L. Collier (eds.) Economics of aquaculture, sea-fishing and coastal resource use in Asia. Agricultural Development Council and Philippine Council for Agriculture and Resources Research, Los Baños, Philippines.

Shumway, S.E., S. Sherman-Caswell and J.W. Hurst. 1988. Paralytic shellfish poisoning in Maine: monitoring a monster. J. Shellfish Res. 7: 643-652.

Taft, W.H. and D.F. Martin. 1986. The potential for managing a Florida red tide. J. Environ. Sci. Health, A21: 107-127.

Taylor, D.L. and H.H. Seliger, editors. 1979. Toxic dinoflagellate blooms. Elsevier North Holland, New York.

Wong, P.S. and R.S.S. Wu. 1987. Red tides in Hong Kong: problems and management strategy with special reference to the mariculture industry. J. Shoreline Manage. 3: 1-21.

Discussion and Recommendations on Economic Issues

Chairperson : J.L. Maclean

Rapporteur : Othman Ross

The discussion group focused on listing the various types of short- and long-term losses and costs potentially or actually associated with *Pyrodinium* red tides and evaluating their relative importance. Rating was done on a scale of 1 (least important) to 5 (most important/severe) in terms of economic impact. The various factors and ratings are given in Table 1.

Positive aspects of red tides were also discussed and a short list was made (Table 2).

The discussion group felt that government agencies dealing with management of *Pyrodinium* red tides situations should carefully evaluate the alternatives suggested by the factors in Table 1 to help them deal systematically with the problem.

In plenary discussions, it was noted that a loss of public confidence in eating seafoods as a result of government mismanagement of information might cost a lot of money over a considerable period to put right.

Also, the cost of items quite unrelated to PSP can increase during red tides scares (e.g., poultry prices) which affects producers and consumers in opposite ways; producers would gain from higher prices of their products at such times.

Finally, because the marketing systems are so complex, a strong recommendation was made to employ economists in order to provide a clear picture of the economic alternatives in red tide management.

Table 1. Actual and potential costs and losses associated with *Pyrodinium* red tides. Numbers represent relative severity of each factor from least (1) to most (5) important.

5	Loss of life
5	Cost of resettlement in a chronic red tide situation (transmigration)
4	Reduced price of uncontaminated seafoods
4	Cost of maintaining public awareness
3	Reduced price of suspect seafood
3	Loss of confidence by consumers
3	Loss of confidence by overseas buyers/markets
3	Loss of income by fishermen, fishfarmers
3	Cost of monitoring and research (personnel, equipment, supplies)
2	Loss of condemned seafood (by either fisherman, fishmonger or middleman)
2	Cost of publicizing and enforcing bans
2	Potential loss of business opportunities (aquaculture, fisheries)
2	Potential loss of resource use
1	Loss of wages of hired labor (fishermen, fishmongers, drivers, etc.)
1	Loss of income by victims, in medicine, time off work, hospitalization (socioeconomic costs)
1	Loss of tourism income (coral reef resources)
1	Loss of income by seafood restaurants
1	Loss of foreign exchange earnings (private and government) (need to import and loss of export)

Table 2. Some possible positive aspects of *Pyrodinium* red tides.

Sector	Item	Comments
Government	Tourism	Possibility of creating phosphorescent bays
Research Institutions	Institution building	Scholarship/grants from foreign assistance agencies for research. Construction and equipment grants. Cooperative research with other institutions
Industry	Soft loans for fishermen	For improving livelihood base
	Harvesting red tides	For extraction of certain chemicals

List of Participants

AUSTRALIA

	:	Dr. Susan Blackburn
	:	Experimental Scientist
	:	CSIRO Marine Laboratories, GPO Box 1538, Hobart, Tasmania 7001, Australia
Telex No.	:	AA 57182
Facsimile No.	:	(002) 240530
Telephone No.	:	(002) 206311
	:	Dr. Gustaaf Marinus Hallegraeff
	:	Principal Research Scientist
	:	CSIRO Marine Laboratories, GPO Box 1538, Hobart, Tasmania 7001, Australia
Telex No.	:	AA 57182
Facsimile No.	:	(002) 240530
Telephone No.	:	(002) 206311

BRUNEI DARUSSALAM

	:	Mr. Matdanan Haji Jaafar
	:	Director of Fisheries
	:	Department of Fisheries, P.O. Box 2161, Bandar Seri Begawan 1921, Brunei Darussalam
Telex No.	:	FISERES BU 2741
Cable No.	:	FISHERIES BRUNEI
Facsimile No.	:	42069
Telephone No.	:	42067/42068/20590
	:	Pg. Sharifuddin Pg Haji Yusof
	:	Deputy Director of Fisheries
	:	Department of Fisheries, P.O. Box 2161, Bandar Seri Begawan 1921, Brunei Darussalam
Telex	:	FISERES BU 2741
Cable No.	:	FISHERIES BRUNEI
Facsimile No.	:	42069
Telephone No.	:	42067/42068/24656

	:	Ms. Siti Amin Mahali
	:	Assistant Fisheries Officer
	:	Fisheries Station Muara, Department of Fisheries, Ministry of Industry and Primary Resources
Telex No.	:	FISERES BU 2741
Cable No.	:	FISHERIES BRUNEI
Facsimile No.	:	90065
Telephone No.	:	90066/90067
	:	Mr. Abdul Halidi Mohd Salleh
	:	Fisheries Officer
	:	Department of Fisheries, P.O. Box 2161, Bandar Seri Begawan 1921, Brunei Darussalam
Telex No.	:	FISERES BU 2741
Cable No.	:	FISHERIES BRUNEI
Facsimile No.	:	42069
Telephone No.	:	42067/42068/90066
	:	Dr. M.W. Ranjith N. De Silva
	:	Fisheries Officer
	:	Department of Fisheries, P.O. Box 2161 Bandar Seri Begawan 1921, Brunei Darussalam
Telex No.	:	FISERES BU 2741
Cable No.	:	FISHERIES BRUNEI
Facsimile No.	:	90065
Telephone No.	:	90066/90067
	:	Mr. Selvanathan Subramaniam
	:	Fisheries Officer
	:	Department of Fisheries, P.O. Box 2161 Bandar Seri Begawan 1921, Brunei Darussalam
Telex No.	:	FISERES BU 2741
Cable No.	:	FISHERIES BRUNEI
Facsimile No.	:	42069
Telephone No.	:	42068/90066
	:	Mr. Abdua Majid Abd Rahman
	:	Principal Biochemist
	:	Chemical Pathology Division, Central State Laboratory, RIPAS Hospital, Bandar Seri Begawan 2680, Brunei Darussalam
Telex No.	:	KKPP BU2813
Cable No.	:	DIRMED BU2369
Telephone No.	:	42424 Ext. 348

	:	Mr. Mohd Amin Abdullah
	:	Chief Health Inspector
	:	Vaccination Centre, RIPAS Hospital, Bandar Seri Begawan 2680, Brunei Darussalam
Telephone No.	:	42424 Ext. 334

	:	Mr. Chong Chee Kiong
	:	Scientific Officer
	:	Ward Pharmacy, RIPAS Hospital, Bandar Seri Begawan 2680, Brunei Darussalam
Telephone No.	:	42424 Ext. 477

	:	Dr. Webber Edric Booth
	:	Senior Lecturer
	:	Biology Department, University Brunei Darussalam, Bandar Seri Begawan, Brunei Darussalam
Telex No.	:	UBD BU 2725
Facsimile No.	:	27003
Telephone No.	:	27001 Ext. 164

	:	Mr. Andrew Murray
	:	Lecturer
	:	Biology Department, University Brunei Darussalam, Bandar Seri Begawan, Brunei Darussalam
Telex No.	:	UBD BU 2725
Facsimile No.	:	27003
Telephone No.	:	27001 Ext. 141

CANADA

	:	Prof. F.J.R. (Max) Taylor
	:	Professor
	:	Dept. of Oceanography, University of British Columbia, Vancouver, B.C., Canada V6T 1W5
Facsimile No.	:	(604) 2286091
Telephone No.	:	(604) 2284587

GUATEMALA

	:	Mr. Fernando Rosales-Loessener
	:	Head of Aquaculture Department
	:	DITEPESCA Ave. Reforma 8-60 Zona 9, Guatemala
Telephone No.	:	313590

ICLARM

	:	Dr. Chua Thia Eng
	:	Director, Coastal Area Management Program, ICLARM
	:	ICLARM, MC P.O. Box 1501, Makati, Metro Manila, Philippines
Telex No.	:	(ITT) 45658 ICLARM PM
Cable No.	:	ICLARM MANILA
Facsimile No.	:	(63-2) 8163183
Telephone No.	:	8180466, 8189283, 8175163

	:	Mr. John Leslie Maclean
	:	Acting Director General
	:	ICLARM, MC P.O. Box 1501, Makati, Metro Manila, Philippines
Telex No.	:	(ITT) 45658 ICLARM PM
Cable No.	:	ICLARM MANILA
Facsimile No.	:	(63-2) 8163183
Telephone No.	:	8180466, 8189283, 8175163

INDONESIA

	:	Ms. Quraisyin Adnan
	:	Researcher
	:	Center for Oceanological Research and Development, National Institute of Science, Jl. Pasir Putih, No. 1, Ancol Timur, P.O. Box 580 DAK Jakarta 11001, Indonesia
Telex No.	:	62875 PDII-IA
Telephone No.	:	683850

	:	Ms. Niniek Suwarni
	:	Administrator
	:	Directorate General of Fisheries, Harsono, R.M. No. 3, Ragunan, Jakarta, Indonesia
Telex No.	:	47318
Telephone No.	:	783226

JAPAN

	:	Dr. Yasuwo Fukuyo
	:	Assistant Professor, University of Tokyo
	:	Yayoi 1-1-1, Bunkyo-ku, Tokyo 113, Japan
Facsimile No.	:	(81-3) 8120529
Telephone No.	:	(81-2) 8122111 Ext. 5291
	:	Dr. Kazumi Matsuoka
	:	Associate Professor, Nagasaki University
	:	Department of Geology, Nagasaki University 1-14 Bunkyo-Machi, Nagasaki, Japan 852
Telephone No.	:	(0958) 471111 Ext. 3274
	:	Dr. Tomotoshi Okaichi
	:	WESTPAC, Leader of Study Group on "Red Tide"
	:	Faculty of Agriculture, Kagawa University, Ikenobe, Miki-cho, Kida-gun, Kagawa Pref, Japan 261-07
Facsimile No.	:	(0878) 987295
Telephone No.	:	(0878) 981411
	:	Dr. Yasukatsu Oshima
	:	Associate Professor, Tohoku University
	:	1-1 Tsutsumi-dori Amemiya, Aoba-ku, Sendai 981, Japan
Telex No.	:	852246 THUCOM J
Facsimile No.	:	(022) 2721870
Telephone No.	:	(022) 2724321

MALAYSIA

	:	Mr. Daim Haji Basrun
	:	Senior Fisheries Officer
	:	Fisheries Department, 8th Floor, Wismakhidmat Building, 88000 Kota Kinabalu, Sabah
Telephone No.	:	52511
	:	Mr. Gires Usup
	:	Senior Assistant Fisheries Officer
	:	UKM Sabah Kampus, Beg Berkunci No. 62, 88996, Kota Kinabalu, Sabah, Malaysia
Telephone No.	:	(088) 211999

	:	Dr. Othman Ross
	:	Lecturer
	:	Faculty of Life Science, Universiti Kebangsaan Malaysia, 43600 Bangi, Selangor, Malaysia
Telex No.	:	MA 31496
Cable No.	:	UNIKEB
Facsimile No.	:	(03) 8256484
Telephone No.	:	(03) 8250001 Ext. 2874/2967

	:	Mr. Ting Thian Ming
	:	Fisheries Officer
	:	Fisheries Department, Kota Kinabalu, Sabah, Malaysia
Telephone No.	:	31322, 35944

PAPUA NEW GUINEA

	:	Mr. Kelly Riroriro
	:	Scientific Officer
	:	P.O. Box 6372, Boroko N.C.D., Papua New Guinea
Telex No.	:	AGRI NE 23076
Facsimile No.	:	214630
Telephone No.	:	217011

PHILIPPINES

	:	Dr. Jose S. Castro
	:	Senior Science Research Specialist
	:	Department of Science and Technology Gen. Santos Avenue, Bicutan, Taguig, Metro Manila, Philippines
Telex No.	:	67608 Access Code 75 (Phil)
Telephone No.	:	8220429, 8220961 Ext. 238/279/277

	:	Dr. Rhodora A. Corrales
	:	Assistant Professor, University of the Philippines
	:	Marine Science Institute, University of the Philippines, Diliman, Quezon City Philippines
Telephone No.	:	982693, 989742

	:	Dr. Cielito Gonzales
	:	Chief, Oceanography Section
	:	Bureau of Fisheries and Aquatic Resources, Quezon Blvd., Quezon City, Philippines
Telephone No.	:	965650
	:	Mrs. Maria Andrea Mendigo
	:	Department of Biochemistry, U.P. College of Medicine, P.O. Box 428 1099 Manila, Philippines
Telephone No.	:	596311, 784831
	:	Mr. Nino Ismael Sanchez Pastor, M.D.
	:	Epidemiologist
	:	Field Epidemiology Training Program, Department of Health, San Lazaro Compound, Sta. Cruz, Manila
Telephone No.	:	7116361

SINGAPORE

	:	Mr. Lim Lian Chuan
	:	Senior Primary Production Officer
	:	Marine Aquaculture Section, Primary Production Department 300 Nicol Drive, Singapore 1749
Telex No.	:	PPD RS 28851
Facsimile No.	:	(065) 5427696
Telephone No.	:	5452124, 5451625

THAILAND

	:	Mr. Kriengsag Saitanu
	:	Instructor
	:	Faculty of Veterinary Science, Chulalongkorn University, Henri Dumont Rd., Bangkok 10500, Thailand
Telephone No.	:	2511656
	:	Dr. Suchana Wisessang
	:	Assistant Professor
	:	Department of Marine Science, Faculty of Science, Chulalongkorn University, Bangkok 10500, Thailand
Telephone No.	:	2516968, 2511951

	:	Ms. Sunee Suvapepun
	:	Chief, Fishery Environmental Research
	:	Marine Fishery Division, 89/1 Sapanpla, Yannawa, Bangkok 10120, Thailand
Telephone No.	:	2114981-2, 2114180

UNITED STATES OF AMERICA

	:	Dr. Donald M. Anderson
	:	Associate Scientist
	:	Biology Department, Woods Hole Oceanographic Institution, Woods Hole, Massachusetts 02543 USA
Telex No.	:	951679
Facsimile No.	:	(1-508) 5481400 Ext. 6108
Telephone No.	:	(1-508) 5481400 Ext. 2351

	:	Dr. Sherwood Hall
	:	Project Manager, Seafood Toxins Research Project
	:	FDA HAFF-423, 200 C Street SW Washington, DC 20204, USA
Telex No.	:	197623 PHS PKLN
Facsimile No.	:	(202) 4261658
Telephone No.	:	(202) 2458418, (202) 2451036

	:	Prof. Howard H. Seliger
	:	Professor
	:	Biology Department, The Johns Hopkins University, 34th and Charles St., Baltimore, Maryland 21218, USA
Telephone No.	:	(301) 3387307

The Participants

Front row: left to right - J.S. Castro, M.D; N. Ismael Sanchez Pastor, M.D.; F. Rosales-Loessener; Dr. Chua Thia-Eng; Abdul Halidi Mohd Salleh; Gires Usup; Maria Andrea Mendigo; Mrs. Niniek Suwarni; Ms. Quraisyin Adnan; Siti Amin Mahali; Ms. Sunee Suvapepun; Rhodora A. Corrales; Hasnah Ibrahim; Rusinah Hj. Yusof; Ms. Suchana Wisessang; Dr. Othman Ross; Matdanan Hj. Jaafar; Prof. Howard H. Heliger; Pg. Sharifuddin Pg. Hj. Yusof; Ramon R. Agbayani.

Second row: left to right - J.L. Maclean; A. Murray; B.J. Pudadera; Mr. Kriengsag Saitanu; Dr. G.M. Hallegraeff; C.L. Gonzales; O.A. Mathisen; Amin Abdullah; Alias Hj. Shari; Chong Chee Kiong; Ajamain Hj. Sawal; Hj. Marzuki Hj. Mohd Salleh; W.S.A.A.L. Kumarasiri; Tatang Sujastani; S. Subramaniam; Ting Thian Ming.

Back row: left to right - Dr. E.W. Booth; Dr. O.M. Anderson; Dr. M.W.R.N. De Silva; K. Riroriro; Lim Lian Chuan; Dr. S. Blackburn; Prof. Tomotoshi Okaichi; Prof. Yasukatsu Oshima.

Jay Maclean (left) and Kelly Riroriro during the opening ceremony.

Sherwood Hall (right) explaining finer points of the mouse bioassay.

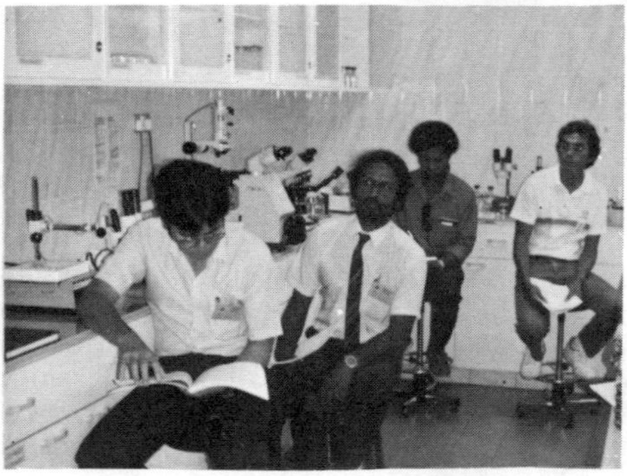

During an introductory lecture on dinoflagellate taxonomy (left and below).

The Brunei Darussalam fish landing complex, site of the laboratory sessions of the workshop.

Attentive participants listening to a lecture on culture methods (left) and comparing notes (below).

Drs. Hallegraeff and Fukuyo discuss a manuscript for the workshop proceedings.

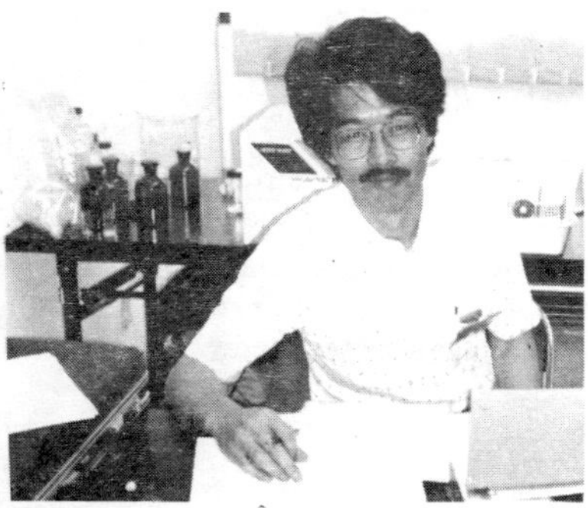

Kazumi Matsuoka.

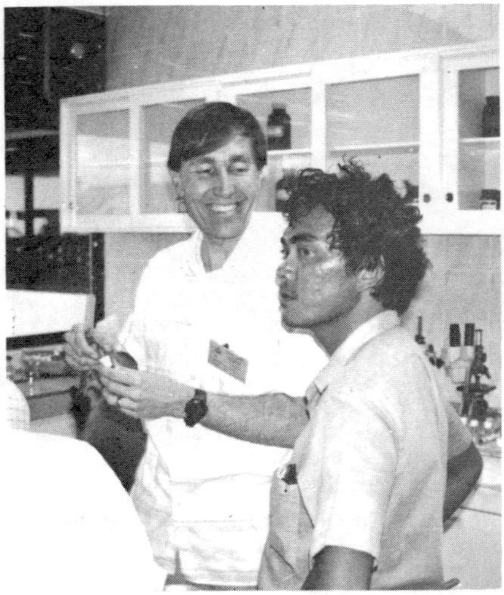

Don Anderson (left) in a light moment with Gires Usup.

Director of Fisheries for Brunei Darussalam, Matdanan Haji Jaafar, donned formal attire to inspect the HPLC equipment. Don Anderson is behind him.

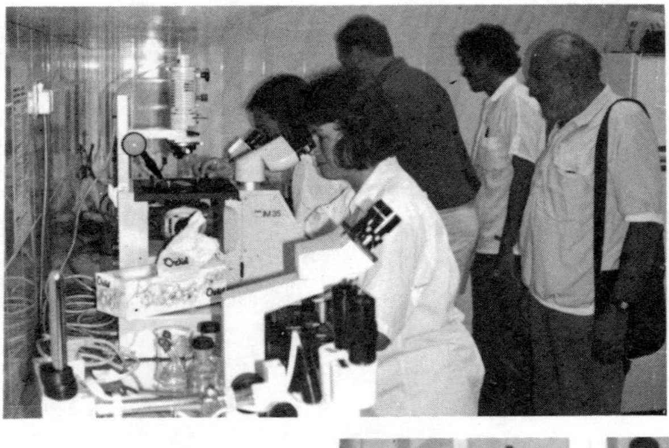

Susan Blackburn (front) and Gustaaf Hallegraeff (back) search for the elusive *Pyrodinium bahamense* cells in Brunei Bay samples. Howard Seliger (right) looks on.

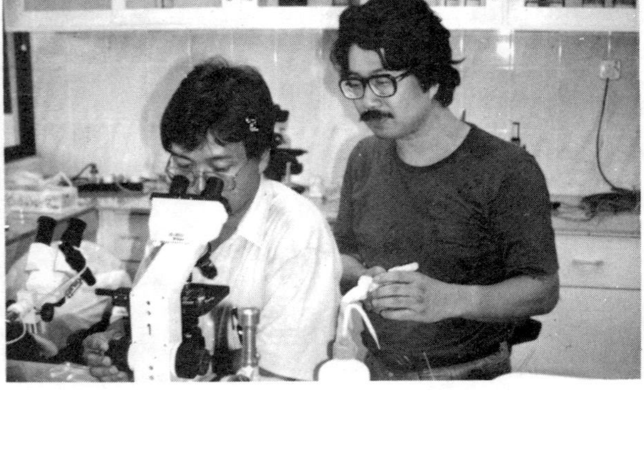

Yasuwo Fukuyo teaching Jose Castro the art of dinoflagellate taxonomy.

Part 2

Manual for Field and Laboratory Research on the Dinoflagellate *Pyrodinium bahamense*

Morphological Characteristics of Dinoflagellates

Y. FUKUYO
Department of Fisheries
University of Tokyo
Yayoi, Bunkyo-ku
Tokyo 113, Japan

F.J.R. TAYLOR
Department of Oceanography and Botany
University of British Columbia
Vancouver, BC, Canada V6T 1W5

FUKUYO, Y. and F.J.R. TAYLOR. 1989. Morphological characteristics of dinoflagellates, p. 201-205. *In* G.M. Hallegraeff and J.L. Maclean (eds.) Biology, epidemiology and management of *Pyrodinium* red tides. ICLARM Conference Proceedings 21, 286 p. Fisheries Department, Ministry of Development, Brunei Darussalam, and International Center for Living Aquatic Resources Management, Manila, Philippines.

Introduction

Dinoflagellates are predominantly unicellular, eukaryotic, flagellated organisms that include both photosynthetic and nonphotosynthetic members. Approximately 2,000 living and 2,000 fossil species have been described. Among the living species only 80 have resting cysts. These are considered to be zygotes (diploid stage) produced during a sexual life cycle. Common planktonic motile forms are cells in a haploid stage, and can reproduce asexually.

Motile Forms

Dinoflagellates have two flagella and are classified into two major groups based on the manner of insertion of the flagella. In the more primitive group, e.g., *Prorocentrum* they are inserted at the anterior end and the cell has no surface grooves.

In the other group, which includes *Pyrodinium,* the two flagella are inserted laterally. One flagellum, termed the longitudinal flagellum, is commonly whip-shaped and orientated posteriorly, starting from an upper sulcal groove (Fig. 1). The other, transverse flagellum, is unique with a helical construction and orientated around the cell in the girdle groove (cingulum). The side from which the flagella arise is ventral, and the opposite is dorsal. Left and right

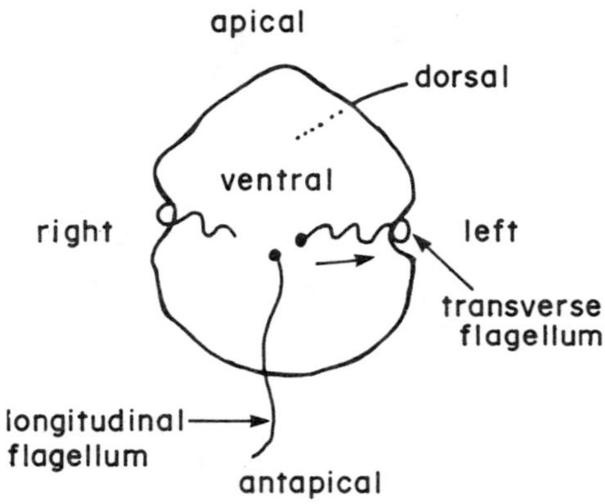

Fig. 1. Orientation.

sides can be recognized by zoological convention. The pole directed towards the swimming direction is apical, and the opposite is antapical. The anterior end of the cell is termed apex, and the opposite is antapex. The girdle essentially divides the cell into an anterior half, the episome (epitheca) and a posterior half, the hyposome (hypotheca). Usually the girdle has displacement, which means that one of the ends of the girdle comes more anterior than the other (Fig. 2). Displacement is said to be left-handed (= descending) or right-handed (ascending) depending on whether the left or right end is more anterior, respectively. The degree of displacement is expressed in girdle widths and is measured from the top edge of the most anterior end to the top edge of the other end.

Depending on rigidity of the cell membrane, this group is commonly subdivided into two groups. The thecate or armored group has a number of cellulosic plates which are called thecal plates, or simply theca as a whole. The other, athecate, unarmored, or naked group lacks the plates. In the thecate group, the thecal plates have

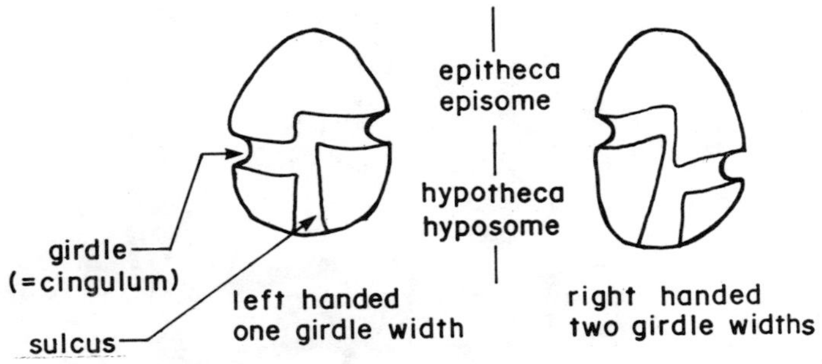

Fig. 2. Girdle displacement.

ornamentations consisting of pores, spines and ridges, making variable, smooth, denticulate, reticulate, or striate markings. The plate junctions are termed sutures. As cells grow in size, thecal plates enlarge at the margins, forming growth bands at sutures. Some species, e.g., *Pyrodinium,* have ridges and spines at sutures.

Athecate species are identified principally by the size and shape of their cells, i.e., cell outline, position of girdle and sulcus groove, girdle displacement. In the thecate group, genera are classified according to plate pattern, which is decided by the arrangement of thecal plates. Species in each genus are identified by the size and shape of the cells, similarly to identification of athecate forms. Therefore, for critical taxonomy of thecate forms, staining or dissociation of thecal plates to recognize their tabulation is necessary in addition to observation of cell shape.

Thecal plates are subdivided into several plate series usually in parallel with the girdle (Fig. 3). All except the sulcal plates are numbered in each series, starting from nearest to the midventral position. Each plate is expressed by the number with the designation of each series (Kofoid notation system).

Apical pore complex (abbreviated as APC): Structures associated with the apical pore at or near apex. A plate bordering the pore is termed Apical pore plate (designated Po), and a plate closing the pore is Apical closing plate (Pi or Pc).

Apical plates (designated '): Plates usually touch the APC.

Precingular plates ("): Plates touch the cingulum on the epitheca.

Anterior intercalary plates (a): Plates between the apicals and precingulars.

Cingular plates (c): Plates in the cingulum.

Postcingular plates ("'): Plates touch the cingulum on the hypotheca.

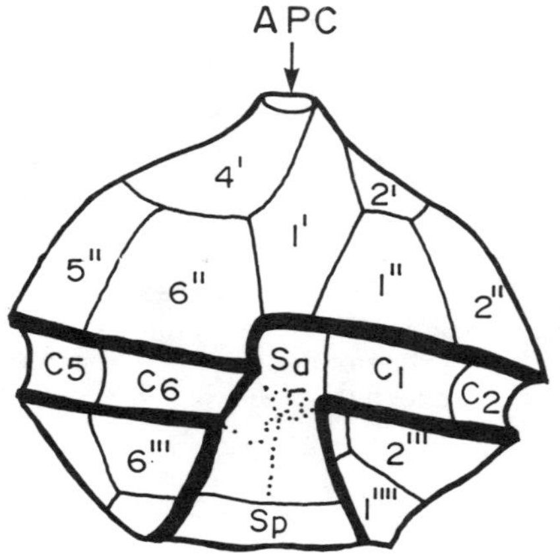

Fig. 3. Plate series notation.

Antapical plates (''''): Plates at the antapex, touching the sulcus.

Sulcal plates (s): Plates in the sulcus. Sulcal plates have their own notation system. For example, sulcal anterior plate is Sa.

In some genera special plate notation, which does not follow the above definition, is permitted to recognize plate homology. In case of *Pyrodinium* the first apical plate (1') does not quite touch the APC, and the first postcingular (1''') is enclosed within the sulcus. If we adopt the definition strictly, the former should be called the first precingular plate (1''), and the latter one of the sulcal plates.

Plate formula is usually used to express the exact number of plates. The formula consists of listing the maximum number of plates in each series, together with their series designation, beginning from the APC to antapicals. *Pyrodinium* can be shown as (Po, Pi), 4', 6'', 6c, 8s, 6''', 2''''. To express the exception in notation of the apical plate series, the formula 3' + 1' sometimes is used instead of the 4'.

Useful References

General taxonomic works:

Drebes, G. 1974. Marine phytoplankton. Thieme, Stuttgard. 186 p.

Dodge, J.D. 1982. Marine dinoflagellates of the British Isles. Her Majesty's Stationery Office, London. 303 p.

Dodge, J.D. 1985. Atlas of dinoflagellates. Farrand Press, London. 119 p.

Hallegraeff, G.M. 1988. Plankton. A Microscopic World. Commonwealth Scientific and Industrial Research Organisation, Hobart, Tasmania. 112 p.

Kofoid, C.A. and O. Swezy. 1921. The free living unarmored dinoflagellata. mem. Univ. Calf. 5: 1-562.

Schiller, J. 1983 and 1937. Dinoflagellate, Rabenhorsts Kryptogamenflora. Leipzig, Vol. 10, Pt. 3, No. 1: 1-617 and No. 2: 1-590.

Sournia, A. 1986. Atlas du Phytoplancton Marin, Vol. 1: Introduction, Cyanophycees, Dictyochophycees, Dinophycees et Raphidophycees. Centre Nat. Rech. Sci., Paris. 219 p.

Steidinger, K. and J. Williams. 1970. Memoirs of hourglass cruises: dinoflagellates. Florida Dept. Nat. Res. 2: 1-251.

Subramanyan, R. 1968. Dinophyceae of the Indian Seas. I. Genus *Ceratium* Schrank. Mar. Biol. Ass. India, Memoir 2. 129 p.

Subramanyan, R. 1971. Dinophyceae of the Indian Seas. II. Family Peridiniaceae Schutt emend Lindemann. Mar. Biol. Ass. India, Memoir 2. 334 p.

Taylor, F.J.R. 1976. Dinoflagellates from the Indian Ocean Expedition. Bibl. Botanica 132: 1-234.

Wood, E.J.F. 1954. Dinoflagellates in the Australian region. Aust. J. Mar. Freshw. Res. 5: 171-357.

Other useful references:

Anderson, D.M., A.W. White and D.G. Baden, editors. 1985. Toxic dinoflagellates. Elsevier North Holland, New York. 561 p.

Dale, B. 1983. Dinoflagellate resting cysts: benthic plankton, p. 69-163. *In* G.A. Fryxell (ed.) Survival strategies of the algae. Cambridge Univ. Press, Cambridge.

Okaichi, T., D.M. Anderson and T. Nemoto, editors. 1989. Red tides: biology, environmental science and toxicology. Elsevier Science Publ., New York. 489 p.

Sarjeant, W.A.S. 1974. Fossil and living dinoflagellates. Academic Press, London. 182 p.

Sournia, A., editor. 1978. Phytoplankton manual. Unesco, Paris. 337 p.

Spector, D.L., editor. 1984. Dinoflagellates. Academic Press, new York. 545 p.

Taylor, D.L. and H.H. Seliger, editors. Toxic dinoflagellate blooms. Elsevier North Holland, New York. 505 p.

Taylor, F.J.R., editor. 1987. The biology of dinoflagellates. Blackwell Sci. Publ., London. 785 p.. .

Morphological Features of the Motile Cell of *Pyrodinium bahamense*

F.J.R. TAYLOR
Department of Oceanography and Botany
University of British Columbia
Vancouver, BC, Canada
V6T 1W5

Y. FUKUYO
Department of Fisheries
University of Tokyo
Yayoi, Bunkyo-ku
Tokyo 113, Japan

TAYLOR, F.J.R. and Y. FUKUYO. 1989. Morphological features of the motile cell of *Pyrodinium bahamense*, p. 207-217. *In* G.M. Hallegraeff and J.L. Maclean (eds.) Biology, epidemiology and management of *Pyrodinium* red tides. ICLARM Conference Proceedings 21, 286 p. Fisheries Department, Ministry of Development, Brunei Darussalam, and International Center for Living Aquatic Resources Management, Manila, Philippines.

Abstract

Descriptions and illustrations of the two varieties of the dinoflagellate *Pyrodinium bahamense* are given. Comparisons with related species are also provided.

The genus *Pyrodinium* is commonly considered to be composed of only one species, *P. bahamense,* although Steidinger et al. (1980) recognized two varieties, var. *bahamense* and var. *compressum,* and Balech (1985) thinks they intergrade completely and do not warrant the separation. Historically, there were seven more species reported, i.e., *P. phoneus, P. schilleri, P. spirale, P. monilatum, P. balechii, P.*

minutum and *P.* sp. All of them are now transferred to other genera and/or synonymized with other species.

Pyrodinium bahamense was first described from New Providence Island of Bahama in the Atlantic Ocean by Plate (1906). In 1931 Böhm reported an anterioposteriorly compressed variation of the species from the Persian Gulf in the Indian Ocean and gave "forma" rank as *P. bahamense* forma *compressa*. The neuter gender of *Pyrodinium* should agree with that of the specific epithet (ICBN 24.2), which thus should be spelled as *compressum*. A similar compressed form was described by Matzenauer (1933) from the Red Sea as *Gonyaulax schilleri*. Schiller (1937) thought *G. schilleri* was a synonym of *P. bahamense* forma *compressum*, and made a new combination, *P. schilleri*. At that time *P. schilleri* was recognized to have differences from *P. bahamense* in its compressed body shape, less pronounced spines, and having one additional thecal plate (four apical plates instead of three). Both *P. bahamense* and *P. schilleri* were reported by Osorio Tafall (1942) from the Pacific coast of Mexico.

Woloszynska and Conrad (1939) described *P. phoneus* from toxic red tides off Belgium, but the species is thought to belong to the genus *Alexandrium* sensu Balech (1985) by its smooth and round cell shape and plate tabulation. A species name *P. spirale* can be found in Margalef (1961), but it could be a typographical error of *Gyrodinium spirale* (Sournia 1973).

The plate pattern of *P. bahamense* was studied in detail again more than two decades after Tafall with special remarks on its life history by Buchanan (1968) and Wall & Dale (1969). Their observations revealed that the pattern was different from the type species of the genus *Gonyaulax,* but coincided with two species of the genus *Gonyaulax,* i.e., *G. monilatum* and *G. balechii,* and one species of the genus *Alexandrium, A. minutum*. Taylor (1976) confirmed the coincidence and transferred these three species to the genus *Pyrodinium* to make the new combinations *P. monilatum, P. balechii* and *P. minutum*. He also described another species as *P.* sp. in the same monograph. Later, however, these three species except *P.* sp. were thought to belong in the genus *Alexandrium* based on differences in characteristics of thecate cells and resting cysts (Steidinger and Tangen 1985).

Steidinger et al. (1980) compared thecate cells of tropical Atlantic and Indo-Pacific *P. bahamense* and recognized several minor morphological and physiological differences which are enough to warrant variety status only. They revised the taxonomy of the species, establishing two varieties, var. *bahamense* for the Atlantic and var. *compressum* for the Indo Pacific material. On the other hand

Balech (1985) also observed morphological variation of both Atlantic and Indo-Pacific specimens independently and concluded that *Pyrodinium* is monospecific and the species, *P. bahamense* cannot be divided into any infraspecific taxa.

Matsuoka et al. (1985) found the resting cyst of the Indo-Pacific variety and reported that the cysts has basically similar morphological characteristics to that of the Atlantic variety described by Wall and Dale (1969) excluding cyst diameter and length of processes on the cyst surface.

Description (Figs. 1, 2, Plates 1, 2 and 3)

Cells single or in chains. Photosynthetic, greenish brown color. Stains darkly in Lugol's Iodine. Single cells almost round, disregarding spines and flanges. Cells in chains apico-antapically flattened. Strongly developed wall (theca) of cellulose plates. Plates ornamented with prominent pores and numerous very fine teethlike pustules/tubercules. Girdle displaced by approximately one girdle width, left handed. Girdle lists strongly developed. Sulcal lists

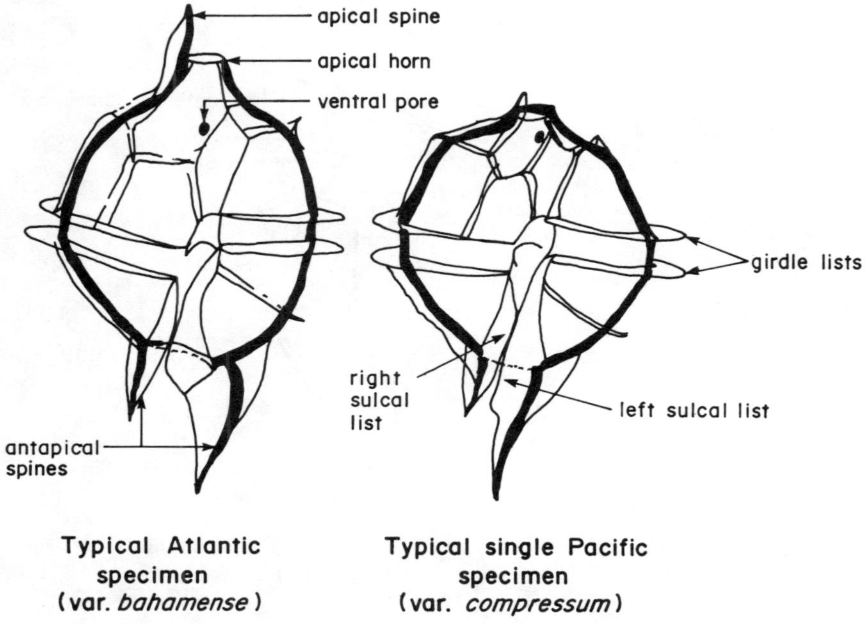

Fig. 1. *Pyrodinium* cell types, based on SEM and Balech (1985).

strongly developed, especially the left, and supported by antapical spines. The left sulcal list encloses the first postcingular plate within the sulcus. Plate pattern is shown as (Po, Pi), 4', 6'', 6c, 8s, 6''', 2''''. Distinctive, prominent flanges arise from most sutures. A well developed ventral pore is present on the fourth apical plate near its margin with the first apical plate. The first apical plate approaches, but does not usually touch the Apical Pore Complex (APC). To express this separation, the formula 3' + 1' sometimes is used instead of the 4'. The APC is triangular and contains two plates: a narrow

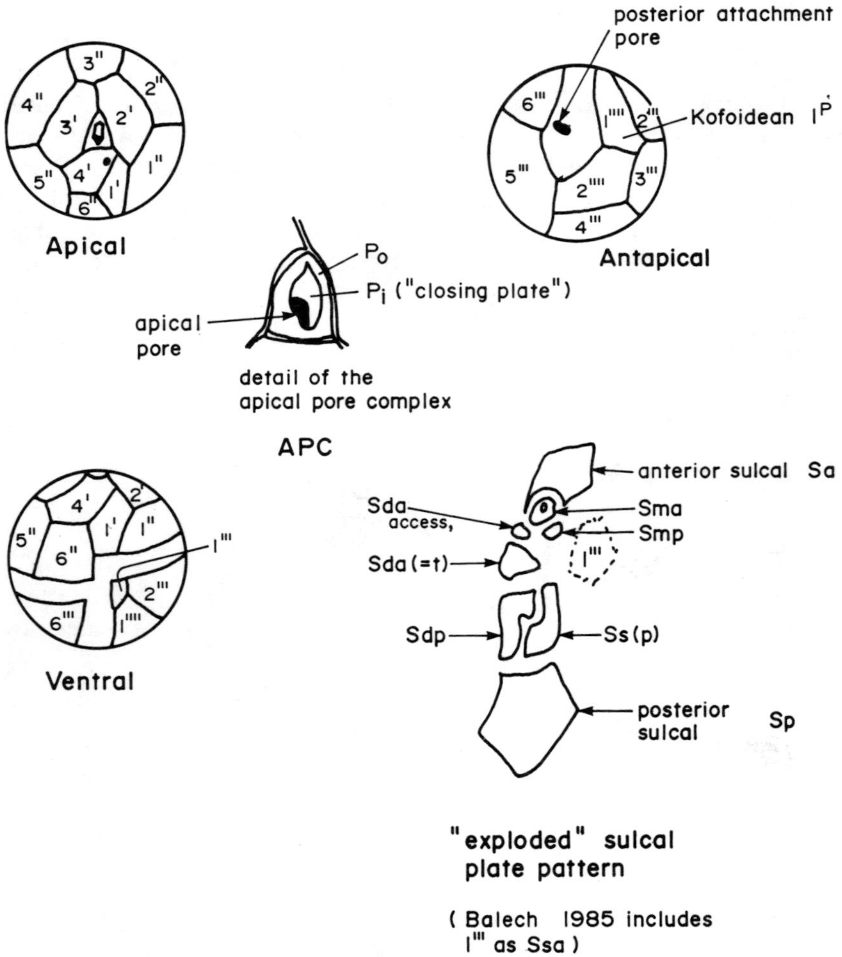

Fig. 2. *Pyrodinium* plate tabulation, using the Kofoid system with Balech modifications (except 1''').

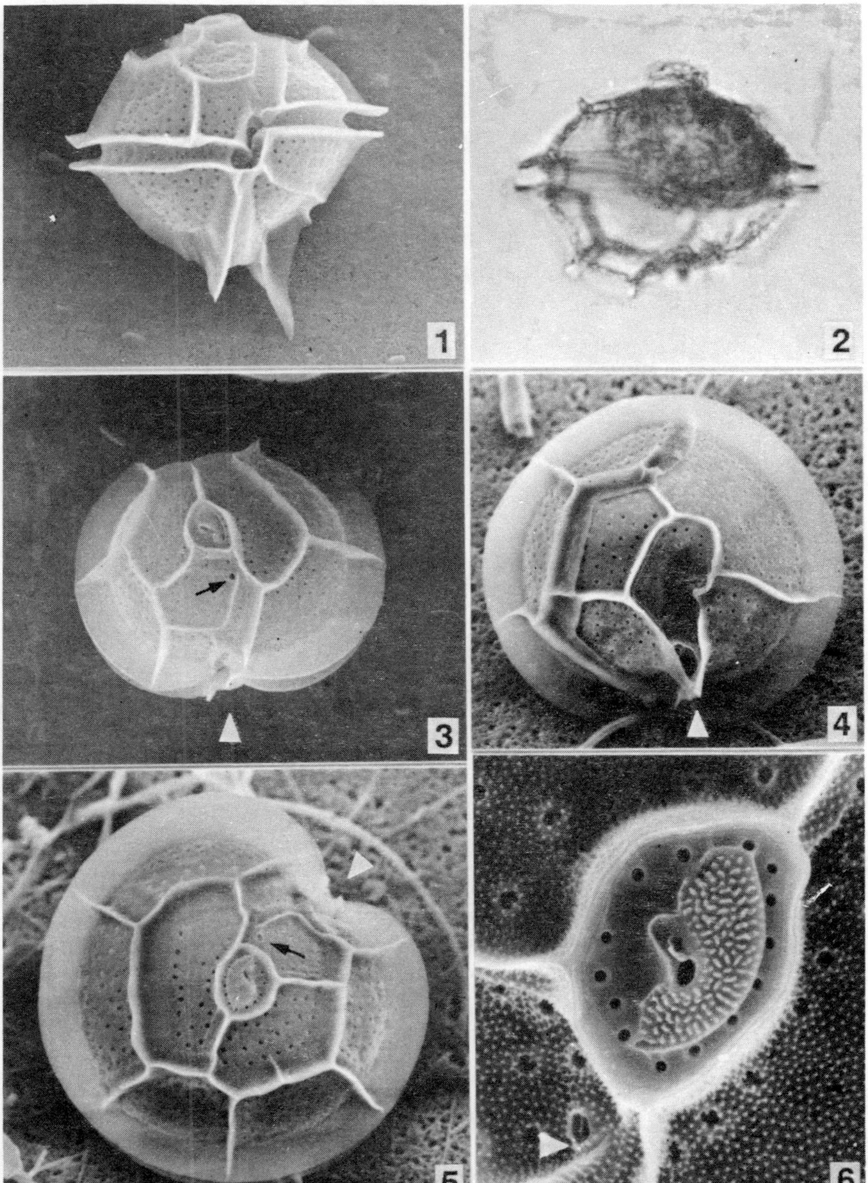

Plate 1. *Pyrodinium bahamense* var. *compressum*

1. ventral view, (SEM)
2. side view (LM)
3. ventro-apical view (SEM), ↑ ventral pore, Δ position of sulcus
4. antapical view (SEM), Δ position of sulcus
5. apical view (SEM), ↑ ventral pore, Δ position of sulcus
6. **Apical Pore Complex (SEM), Δ ventral pore.**

Plate 2. Scanning micrographs of *Pyrodinium bahamense* thecae

1. Var. *bahamense* from Oyster Bay (Jamaica) with well developed apical horn (Ah) and spine (Sp). A ventral pore (Vp) is present in 4'. Scale bar 10 μm.
2. Var. *bahamense.* Monospecific bloom sample from Oyster Bay containing single cells only. Scale bar 100 μm.
3. Var. *compressum* from Papua New Guinea. Anterio-posteriorly compressed cell with well-developed antapical spine (As) and left sulcal list (LsL). Scale bar 10 μm.
4. Cell from Papua New Guinea, Intermediate between var. *compressum* and var. *bahamense.* The epitheca has 1 apical and 6 precingular plates.
5. Var. *compressum* four-celled chain from the Philippines.
6. Apical view of cell from Papua new Guinea showing the apical pore complex (Po) surrounded by 4 apical plates (Po and 1' not in contact).
7. Antapical view of cell from Papua New Guinea showing the posterior attachment pore (Pp) and 2 postcingular plates.

(Micrographs: G.M. Hallegraeff, except Fig. 5, Y. Fukuyo; Samples: R. Buchanan (Figs. 1, 2), J.L. Maclean (Figs. 3, 4, 6, 7), Y. Fukuyo (Fig. 5)).

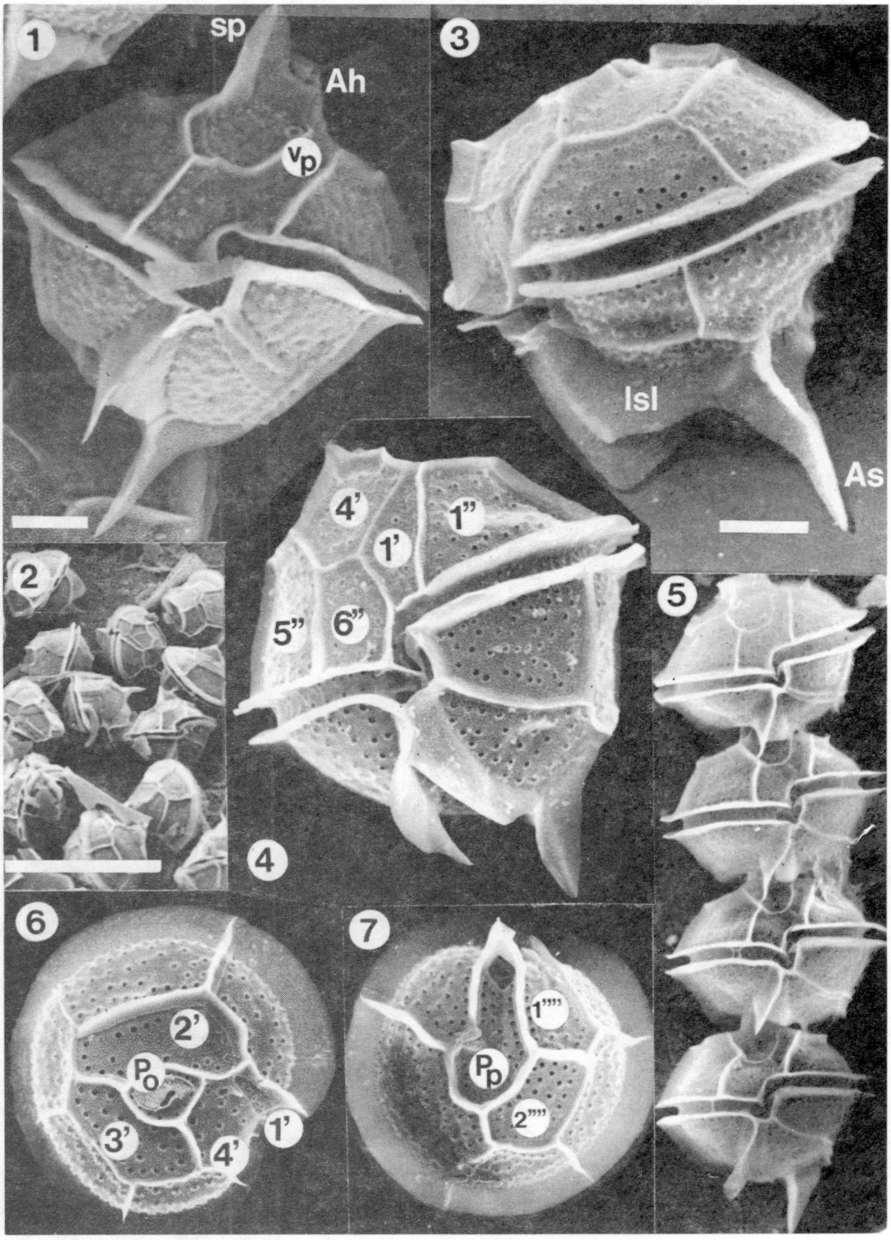

Plate 2.

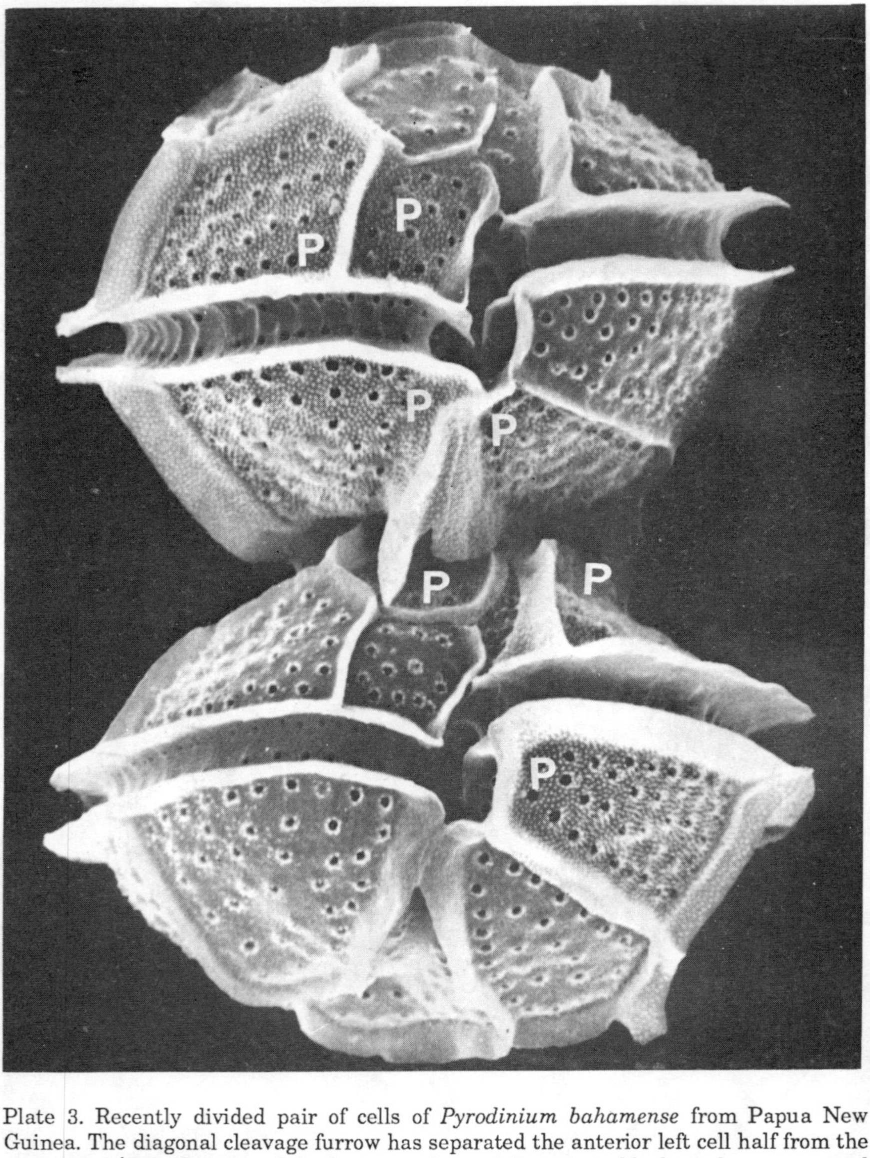

Plate 3. Recently divided pair of cells of *Pyrodinium bahamense* from Papua New Guinea. The diagonal cleavage furrow has separated the anterior left cell half from the posterior right cell half, with both parent thecae (p) recognizable from the ornamented thecal plates.

The anterior daughter cell has regenerated a new right half and the posterior daughter cell has regenerated a new left half, both with smooth thecal plates.

(Micrograph G.M. Hallegraeff; Sample J.L. Maclean)

outer (Po) and a leaf-shaped inner (Pi) plate with a narrow, slit-like apical pore on its margin.

Atlantic populations (Tampa Bay, Florida and Oyster Bay, Jamaica), *P. bahamense* var. *bahamense,* may occur in pairs but do not usually form chains. Their apical spines are usually more developed. Length and transdiameter are 33-71 and 33-67 μm, respectively.

Indo-Pacific populations (Papua New Guinea, Sabah and Brunei), *P. bahamense* var. *compressum,* are anterio-posteriorly compressed and may make long chains of up to thirty-two cells. They occasionally produce paralytic shellfish toxins. Length, transdiameter and dorso-ventral diamter in Brunei are 33-45, 37-47 and 37-47 μm, respectively. In Sabah, 33-47, 39-52 and 39-47 μm, respectively.

Comparisons

P. bahamense is fortunately very distinctive. Within the region where this species is found it might be confused with *Triadinium (= Goniodoma) polyedricum* which also has flanges along its sutures (see below). However, the latter is more angular in shape, never occurs in chains, and lacks the prominent spines of *Pyrodinium.* The APC of the latter is not separable into Po and Pc, but that of the former is composed of the two platelets.

Plate arrangement of *Pyrodinium* is virtually identical with some *Alexandrium* species such as *A. minutum, A. monilatum* and *A. pseudogonyaulax.* But the *Alexandrium* spp. have spherical shape with rounded profile. Their thecal plates are much thinner, delicate and smooth.

Some species of *Alexandrium,* i.e., *A. catenella, A. cohorticula* and *Gymnodinium catenatum* form long chains of cells. Although the chains look similar to that of *Pyrodinium* under low magnification, the cells in the chains are round, not polygonal as in *Pyrodinium. Alexandrium* has delicate thecal plates and *Gymnodinium* lacks the plates. As *Pyrodinium,* however, has thick thecal plates, it is useful to add a drop of sodium hypochlorite solution to dissociate the plates from the cell surface. Some species of *Gonyaulax,* for example, *Gonyaulax (Lingulodinium) polyedra,* may also be confused but these occur singly, lack flanges and have a greater girdle displacement.

Morphological Characteristics of *Triadinium polyedricum* (Plate 4)

Cells single, never making chains. Photosynthetic, brown color. Cells polygonal, seven-sided in ventral view, epitheca with three

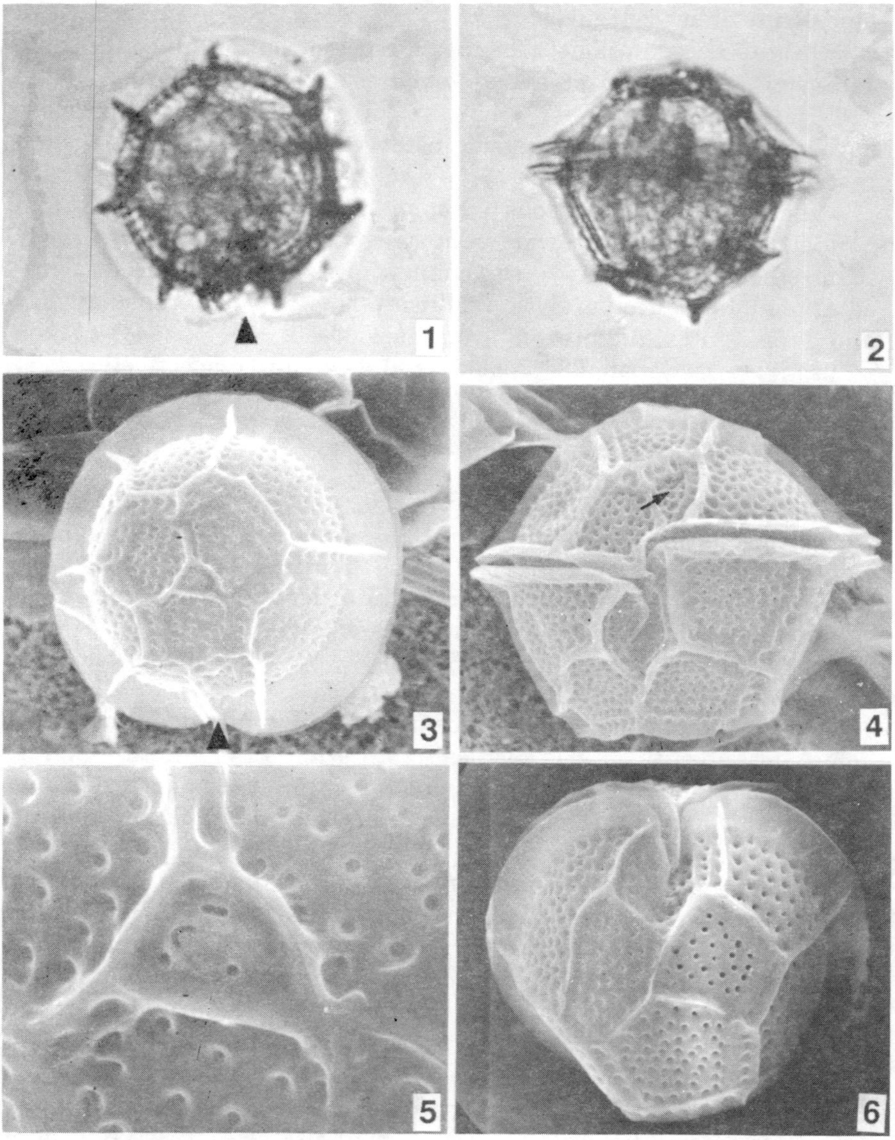

Plate 4. *Triadinium polyedricum*

1. apical view (LM), Δ position of sulcus
2. side view (LM)
3 apical view (SEM), Δ position of sulcus
4. ventral view, ↑ ventral pore
5. Apical Pore Complex (SEM)
6. ventro-antapical view (SEM)

corners, hypotheca with two. Girdle equatorial, slightly left-handed, with wide lists. Sulcus short and broad. Plate pattern is Po, 3', 7", 6c, 6s, 5"', 3"". Thecal plates ornamented with pores and with strong ridges arising from the sutures. A small ventral pore is present on the first precingular plate near its margin with the first apical plate. The APC triangular and surrounded by three apical plates of similar size and shape. Two platelets of APC, outer (Po) and inner (Pi), fuse into one plate.

References

Balech, E. 1985. A redescription of *Pyrodinium bahamense.* Plate (Dinoflagellata). Rev. Palaeobot. Palyn. 45: 17-34.

Böhm, A. 1931. Peridineen aus dem Persischen Golf und dem Golf von Oman. Arch. protistenkd. 74: 188-197.

Buchanan, R.J. 1968. Studies at Oyster Bay in Jamaica, West Indies, IV. Observations on the morphology and asexual cycle of *Pyrodinium bahamense* Plate. J. Phycol. 4: 272-277.

Margalef, R. 1961. Hidrografia y fitoplancton de un area marina de la cost meridional de Puerto Rico. Inv. Pesq. 18: 33-96.

Matsuoka, K., Y. Fukuyo and C.L. Gonzales. 1985. A new discovery of cyst of *Pyrodinium bahamense* var. *compressum* from the Samar Sea, Philippines, p. 301-304. *In* T. Okaichi, D.M. Anderson and T. Nemoto (eds.) Red tides, biology, environmental science and toxicology. Elsevier Science Publ., New York.

Matzenauer, L. 1933. Die Dinoflagellaten des Indischen Ozeans. (Mit Ausnahme der Gattung *Ceratium*). Bot. Arch. 35(4): 437-510.

Osorio Tafall, B.F. 1942. Notas sobre algunos dinoflagelados planctonicos marinos de Mexico, con descripcion de nuevas especies. An. E. N. Cienc. Biol. 2: 435-447.

Plate, L. 1906. *Pyrodinium bahamense* n.g., n. sp. die Leucht-Peridinee des "Feuersees" von Nassau, Bahamas. Arch Protistenkd. 7: 411-429.

Schiller, J. 1937. Dinoflagellatae, Rabenhorsts Kryptogamenflora. Leipzig 10(3) No. 2: 1-590.

Sournia, A. 1973. Catalogue des especes et taxons infraspecifiques de Dinoflagelles marins actuels, I. Dinoflagelles Libres. Beihefte Nova Hedwigia 48. 92 p.

Steidinger, K.A. and K. Tangen. 1985. Taxonomy and systematics, p. 534-537. *In* D.M. Anderson, A.W. White and D.G. Baden (eds.) Toxic dinoflagellates. Elsevier North Holland, New York.

Steidinger, K.A., L.S. Tester and F.J.R. Taylor. 1980. A redescription of *Pyrodinium bahamense* var. *compressa* (Böhm) stat. nov. from Pacific red tides. Phycologia 19(4): 329-337.

Taylor, F.J.R. 1976. Dinoflagellates from the Indian Ocean Expedition. Bibl. Botanica 132: 1-234.

Wall, D. and B. Dale. 1969. The "hystrichosphaerid" resting spore of the dinoflagellate *Pyrodinium bahamense* Plate 1906. J. Phycol. 5: 140-149.

Woloszynska, J. and W. Conrad. 1939. *Pyrodinium phoneus,* n.sp., agent de la toxicite des moules de canal maritime de Bruges a Zeebrugge. Bull. Mus. R. Hist. Nat. Belgique 15: 1-5.

Morphological Features of the Cyst of *Pyrodinium bahamense* var. *compressum*

K. MATSUOKA
Department of Geology
Faculty of Liberal Arts
Nagasaki University
1-14 Bynkyo-Machi
Nagasaki, 852 Japan

MATSUOKA, K. 1989. Morphological features of the cyst of *Pyrodinium bahamense* var. *compressum*, p. 219-229. *In* G.M. Hallegraeff and J.L. Maclean (eds.) Biology, epidemiology and management of *Pyrodinium* red tides. ICLARM Conference Proceedings 21, 286 p. Fisheries Department, Ministry of Development, Brunei Darussalam, and International Center for Living Aquatic Resources Management, Manila, Philippines.

Abstract

Cysts of *Pyrodinium bahamense* var. *compressum* and several similar species are compared with special attention to differences in process characters. Keys to these species with and without reference to archeopyle features are provided.

Introduction

Cysts of dinoflagellate red tide organisms are regarded as a seed population for a forthcoming bloom. For predicting red tides, plankton researchers and fisheries scientists should have an accurate knowledge of the geographical distribution of cysts. However, since the cyst morphology is quite different from that of motile cells and many similar cysts are present in modern surface sediments, the observation and identification of dinoflagellate cysts are difficult. Additionally, some modern cysts have two different names, because the cyst classification system developed by paleopalynologists is independent of that of motile dinoflagellate cells.

This article documents the morphological differences among the cysts of *Pyrodinium bahamense* var. *compressum* and similar taxa.

Taxonomical Note on the Cyst of *Pyrodinium bahamense* var. *compressum* and Related Taxa

Pyrodinium bahamense Plate var. *compressum* (Böhm), thecate cells have been recorded in several places in the Asian tropical region (Beales 1976; Steidinger et al. 1980; Harada et al. 1982; Hermes and Villoso 1983; Maclean 1984), but only one record of the occurrence of its cyst has been made in surface sediments in the region (Matsuoka et al. 1989).

The surface sediments in the Samar Sea of the Philippines contain abundant cysts of *Pyrodinium bahamense* var. *compressum* (= *Polysphaeridium zoharyi* (Rossignol) Bujak et al. in paleontological systematics), *Gonyaulax* spp. (= *Spiniferites* spp.), *Protoceratium reticulatum* Claparede and Lachmann (senior synonym of *Gonyaulax grindleyi* Reinecke) (= *Operculodinium centrocarpum* (Deflandre and Cookson) Wall and possibly *O. israelianum* (Rossignol) Wall, *Lingulodinium polyedra* (Stein) Dodge (= *Lingulodinium machaerophorum* (Deflandre and Cookson) Wall) and others (Matsuoka, unpublished data). Among them, the cysts of *Lingulodinium polyedra* and *Protoceratium reticulatum* are quite similar to the cyst of *Pyrodinium bahamense* var. *bahamense* and *P. bahamense* var. *compressum* in morphology. The occurrence of these similar cysts is expected in other subtropical and tropical regions in Southeast Asia.

In surface sediments, there are two types of cysts, namely living cysts which are filled with protoplasm and may germinate after a considerable resting period, and empty cysts which have already hatched and bear an excystment aperture, the archeopyle. At present the archeopyle type is one of the most important characters for the identification of dinoflagellate cysts. However, to produce a strain of a certain dinoflagellate species from a cyst, we need the living cyst before germination. In that case, the cyst must be observed and identified for experiments without any information concerning the archeopyle type. Terminology for cyst processes follows Sarjeant (1982).

***Pyrodinium bahamense* var. *bahamense* and *P. bahamense* var. *compressum* *Polysphaeridium zoharyi* (Rossignol) Bujak et al. (Figs. 1-4; Fig. 8A; Figs. 10-13)**

Spherical to ovoidal cyst (50-70 µm in diameter; Fig. 3); wall colorless and composed of two layers, periphragm and endophragm adpressed between processes; periphragm coarsely granulate (Fig. 4), and forming processes (13-18 µm in length), which are hollow,

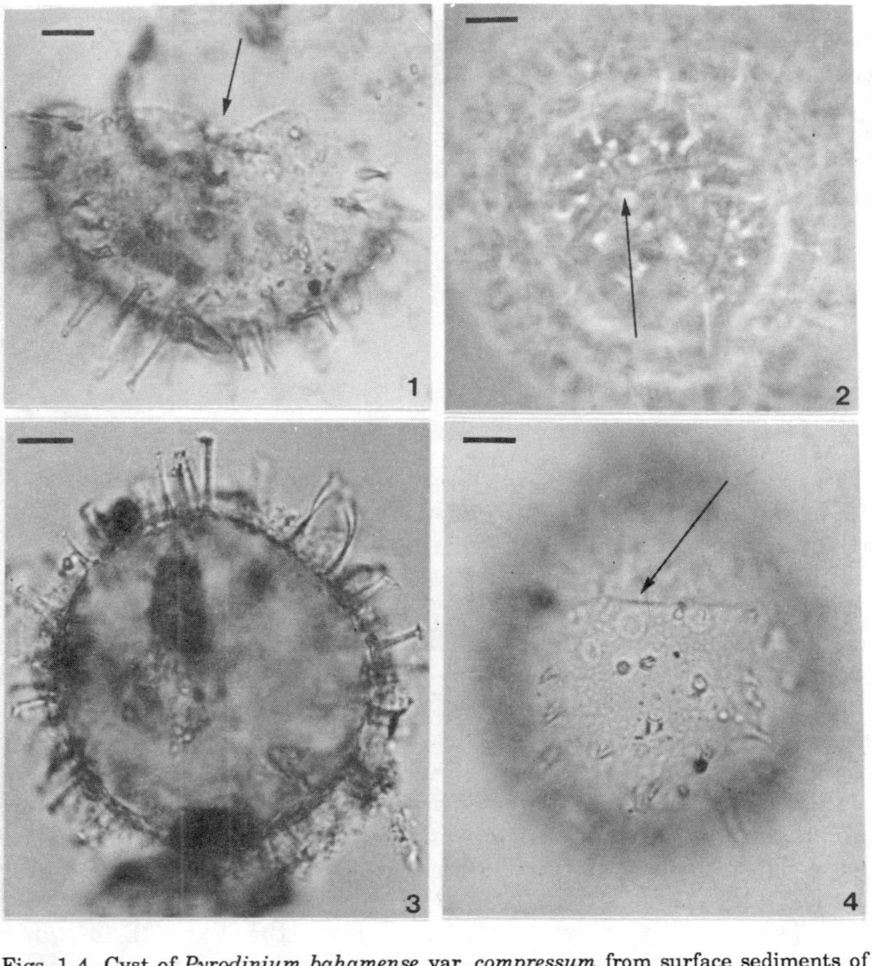

Figs. 1-4. Cyst of *Pyrodinium bahamense* var. *compressum* from surface sediments of the Samar Sea, Philippines: 1, Hypocyst in ventral view, arrow showing sulcal notch; 2, cyst surface, arrow showing archeopyle sutures (phase contrast); 3, optical cross section of dorso-ventral view, showing long, slender tubiform processes; 4, coarsely granulate cyst surface, arrow showing archeopyle sutures.

intratabular, slender and cylindrical to tubiform with capitate or aculeate and open distal extremities (Fig. 3, Fig. 8A); archeopyle basically epicystal (Fig. 1), but the development of archeopyle sutures variable and then sometimes seemingly of a combination precingular type (Fig. 2); after germination, cyst sometimes spread into two hemispherical parts on the major archeopyle suture (Fig. 1).

There are no considerable morphological differences between the cysts of *P. bahamense* var. *bahamense* and *P. bahamense* var. *compressum* except for cyst diameter and the length of processes.

Two subspecies are recognized in the fossil form: *Polysphaeridium zoharyi* subsp. *zoharyi* (Rossignol) Bujak et al. which is characterized by long processes and *Polysphaeridium zoharyi* subsp. *ktana* (Rossignol) Lentin and Williams possessing relatively short processes. The fossil cyst was first described from the Pleistocene sediments of Israel by Rossignol (1964) and later Wall and Dale (1968) confirmed its thecate form to be *Pyrodinium bahamense* var. *bahamense* on the basis of a unialgal cyst culture from the Caribbean Sea. Several fossil locations where *Polysphaeridium zoharyi* occur around the Pacific include Central Kinki in Japan (Matsuoka 1976; Harada 1986) and northern New South Wales in Australia (McMinn 1987, 1989) (see map in Fig. 9)

Lingulodinium (Gonyaulax) polyedra (Stein) Dodge *(Lingulodinium machaerophorum* (Deflandre and Cookson) Wall) (Fig. 5, Fig. 8D)

Spherical cyst (45-60 μm in diameter); wall colorless and comprising two adpressed layers, a granulate periphragm and thin endophragm; processes hollow, nontabular, flexous, acuminate, evexate and/or bulbous with closed distal extremities (12-30 μm in length, Fig. 8D); archeopyle precingular combination consisting three to five precingular paraplates or precingular and anterior intercalary combination comprising four precingular and two anterior intercalary paraplates; the cyst after excystment being hemispherical with zigzag archeopyle sutures.

This cyst is widely distributed from the tropical to warm temperate climatic zone. It was first described from Australian Miocene sediments by Deflandre and Cookson (1955) and later Wall and Dale (1968) clarified its thecate form *(Lingulodinium polyedra)* from a cyst incubation experiment.

Protoceratium reticulatum Claparede and Lachmann - *Type 1* (*Operculodinium centrocarpum* (Deflandre and Cookson) Wall) (Figs. 6, 7; Fig. 8B)

Spherical cyst (30-40 μm in diameter; Fig. 6; Fig. 8B); wall colorless and consisting of two adpressed layers, the periphragm and endophragm; outer surface of periphragm granulate; processes hollow, nontabular, and slender, long, and cylindrical to buccinate with capitate and closed distal extremities (up to 10 μm in length,

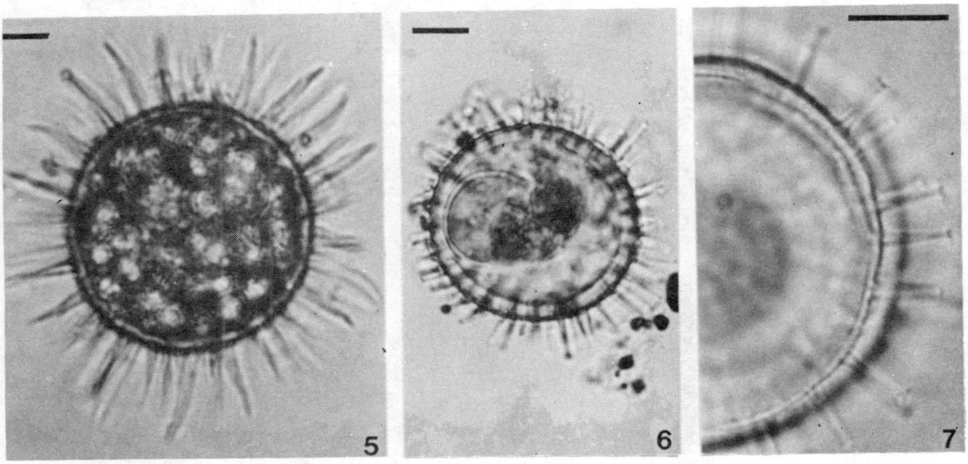

Fig. 5. Living cyst of *Lingulodinium polyedra* in optical cross section, showing acuminate processes.
Figs. 6-7. Living cyst of *Protoceratium reticulatum* Claparéde and Lachmann: 6, optical cross section; 7, Cylindrical to tubiform processes with capitate distal extremities. Scale bar is 10 μm.

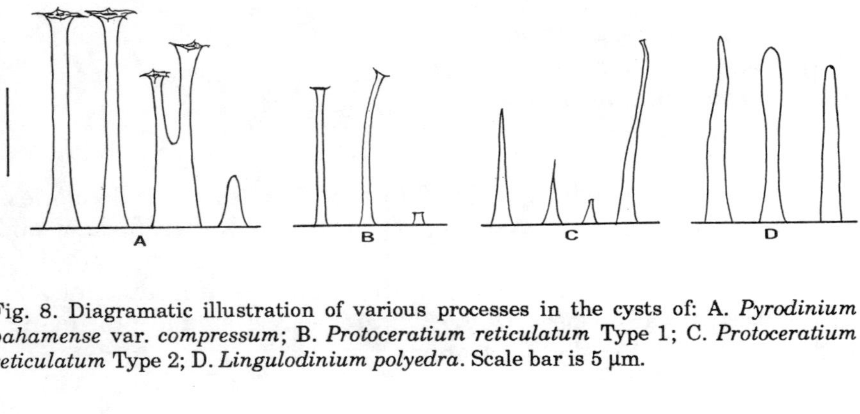

Fig. 8. Diagramatic illustration of various processes in the cysts of: A. *Pyrodinium bahamense* var. *compressum*; B. *Protoceratium reticulatum* Type 1; C. *Protoceratium reticulatum* Type 2; D. *Lingulodinium polyedra*. Scale bar is 5 μm.

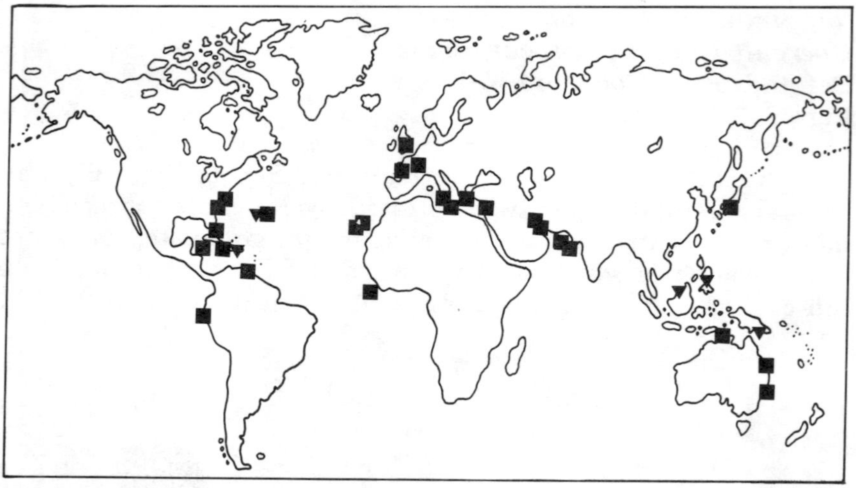

Fig. 9. Global distribution of fossil (squares) and recent (triangles) *Pyrodinium bahamense* cysts (compiled by G.M. Hallegraeff, B. Dale and K. Matsuoka).

Fig. 7) or rarely very short nodular (less than 3 μm in length; Fig. 8B); archeopyle simple precingular, and formed by loss of the 3" paraplate.

This cyst is cosmopolitan and occurs widely from the tropical to subarctic climatic zones. The fossil cyst was first described from Miocene sediments of Australia by Deflandre and Cookson (1955). Later its thecate form was confirmed in a unialgal incubation carried out by Wall and Dale (1968). However, some morphological differences between the Miocene and modern cysts have been observed and further detailed investigation is required.

***Protoceratium reticulatum* Claparede and Lachmann - Type 2 (*Operculodinium israelianum* (Rossignol) Wall) (Fig. 8C)**

Spherical to ovoidal cyst (60-75 μm in diameter); wall colorless and comprising two adpressed layers, a granulate periphragm and thin endophragm; processes densely distributed, short conical or acuminate (5-20 μm in length) with spinate or rarely truncated extremities distally closed (Fig. 8C); archeopyle simple precingular and formed by loss of the 3" paraplate.

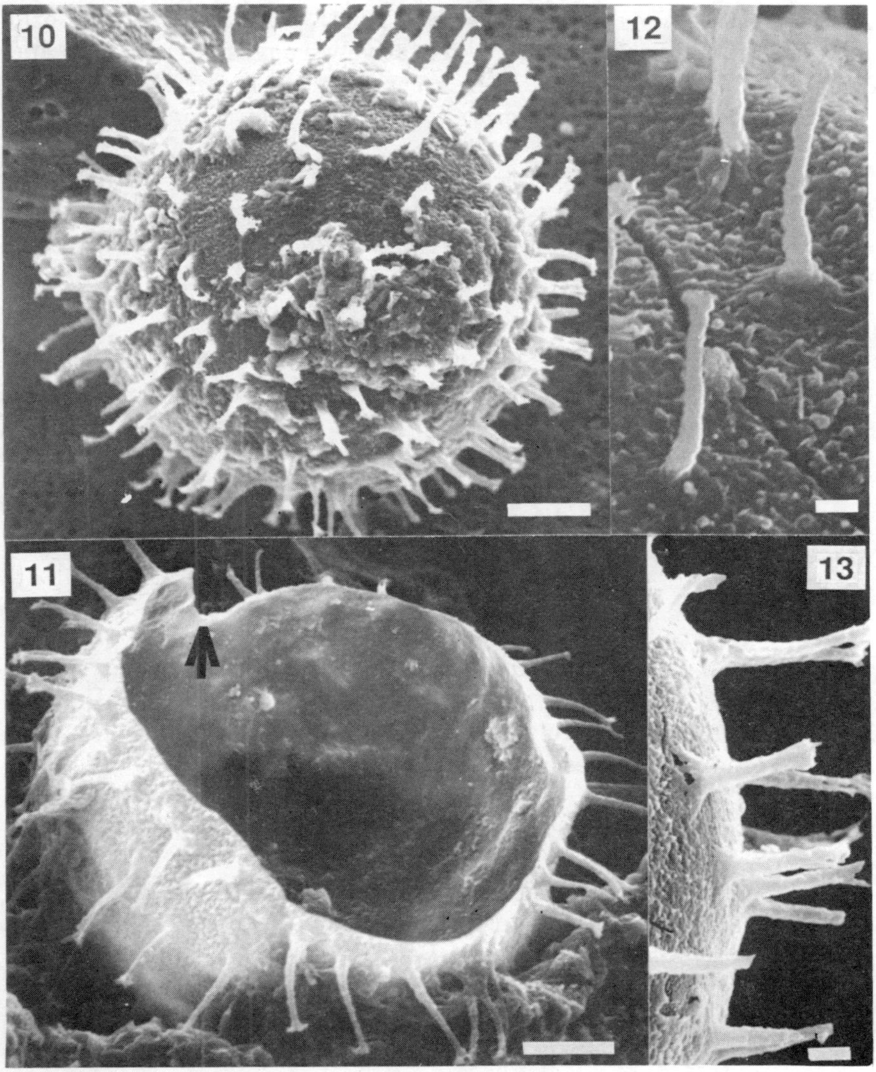

Figs. 10-13. Scanning electron micrographs of *Pyrodinium bahamense* cysts from Port Moresby Harbor, Papua New Guinea.

10. Intact, spherical cyst with numerous radiating spines.
11. Germinated cyst which has opened up equatorially to form two hemispheres (epicystal archeopyle). The remaining hypocyst shows a characteristic sulcal notch (arrow).
12. Detail of microgranular cyst surface.
13. Detail of tubular spines with capitate tips.

(Micrographs: G.M. Hallegraeff; Sample: J.L. Maclean)
Scale bars 10 μm (Figs. 10, 11), 1 μm (Figs. 12, 13)

This cyst was originally described from the Pleistocene sediments of Israel by Rossignol (1964) and its thecate form *Protoceratium reticulatum* was tentatively suggested by Wall and Dale (1968). This cyst has been recorded from the modern surface sediments in the warm temperate to tropical climatic zone.

The morphological characters of the cysts of *Pyrodinium bahamense* var. *bahamense* and *P. bahamense* var. *compressum, Lingulodinium polyedra* and *Protoceratium reticulatum* can be summarized as follows:

Common characters: spherical to ovoidal shape, colorless and granulate wall comprising two adpressed layers, and numerous spinose processes;

Different characters: archeopyle type and process forms.

Key to Species with Spherical Cyst Body and with Numerous Processes

1. Brownish wall consisting of autophragm with many long acuminate processes

--- *Diplopelta parva*

1. Colorless wall composed of two adpressed layers with numerous spines --- *2*

2. Long, slender, cylindrical processes with capitate or aculeate distal extremities --- *3*

2. Short conical or acuminate processes with spinate or truncate distal extremities

--- *Protoceratium reticulatum* *(Operculodinium israelianum)*

2. Long flexous and acuminate, evexate or bulbous processes

--- *Lingulodinium polyedra* *(Lingulodinium machaerophorum)*

3. Granulate cyst surface with smaller body (less than 40 µm in diameter)

--- *Protoceratium reticulatum* (*Operculodinium centrocarpum*)

3. Coarsely granulate cyst surface with larger body (more than 50 µm to less than 70 µm in diameter)

--- *Pyrodinium bahamense* var. *bahamense* and *P. bahamense* var. *compressum*

From the above key, when there is no information concerning the archeopyle (as in living cysts), the cyst of *P. bahamense* var. *bahamense* and *P. bahamense* var. *compressum* is distinguishable from *L. polyedra* and *Protoceratium reticulatum* Type 2 (*Operculodinium israelianum*) in having long cylindrical processes, and from *Protoceratium reticulatum* Type 1 (*Operculodinium centrocarpum*) in possessing a larger cyst body and coarser cyst surface.

When archeopyle features can be distinguished, the identification of *Pyrodinium bahamense* cysts is more straightforward.

Key to spinose cyst species with spherical body and archeopyle features visible.

1. Therophylic archeopyle (operculum adnate)

--- *Diplopelta parva*

1. Saphophylic archeopyle (operculum free) --- *2*

2. Colorless cyst wall with epicystal or combination precingular archeopyle (operculum compound) --- *3*

2. Colorless cyst wall with simple precingular archeopyle (3" paraplate --- *4*

3. Archeopyle with zigzag sutures and composed of 3-5 precingular or 4 precingular + 2 anterior intercalary paraplates

--- *Lingulodinium polyedra* (*Lingulodinium machaerophorum*)

3. Archeopyle epicystal with conspicuous sulcal notch

--- *Pyrodinium bahamense*
(Polysphaeridium zoharyi)

4. Cyst diameter 30-40 µm

--- *Protoceratium reticulatum*
(Operculodinium centrocarpum)

4. Cyst diameter 60-75 µm

--- *Protoceratium reticulatum*
(Operculodinium israelianum)

Acknowledgements

I am grateful to Drs. T. Nemoto and T. Okaichi for giving an opportunity to attend the Management and Training Workshop on *Pyrodinium* Red Tide in Bandar Seri Begawan, Brunei Darussalam, and also Drs. Y. Fukuyo, G.M. Hallegraeff and Mr. J.L. Maclean for their fruitful discussion concerning *Pyrodinium* cysts.

References

Beales, R.W. 1976. A red tide in Brunei waters. Brunei Mus. J. 3(4): 167-182.

Deflandre, G. and I.C. Cookson. 1955. Fossil microplankton from Australian late Mesozoic and Tertiary sediments. Aust. J. Mar. Freshw. Res. 6: 242-313.

Dodge, J.D. 1983. Some revisions of the family *Gonyaulacaceae* (Dinophyceae) based on a scanning electron microscope study. Botanica Marina 32: 275-298.

Harada, K. 1984. Study of dinoflagellates from the submarine strata at the Kansai International Airport in Osaka Bay off Senshu, Central Japan, p. 77-90. *In* Geological survey of the submarine strata at the Kansai International Airport in Osaka Bay, Central Japan. Report of the Calamity Science Institute, Osaka, Japan.

Harada, T., Y. Oshima, H. Kamiya and T. Yasumoto. 1982. Confirmation of paralytic shellfish toxins in the dinoflagellate *Pyrodinium bahamense* var. *compressa* and bivalves in Palau. Bull. Japan Soc. Sci. Fish. 48(6): 821-825.

Hermes, R. and E.P. Villoso. 1983. A recent bloom of the toxic dinoflagellate *Pyrodinium bahamense* var. *compressa* in central Philippine waters. Fish. Res. J. Philipp. 8(2): 1-8.

Maclean, J.L. 1984. Indo-Pacific red tides, p. 173-178. *In* D.L. Taylor and H.H. Seliger (eds.) Toxic dinoflagellate blooms. Elsevier, North Holland.

Matsuoka, K. 1976. Paleoenvironmental study of the Saho and Saidaiji Formations from a view point of palynology. Bull. Mizunami Fossil Mus. 3: 99-117.

Matsuoka, K., Y. Fukuyo and C.L. Gonzales. 1989. A new discovery of cysts of *Pyrodinium bahamense* var. *compressum* from the Samar Sea, Philippines, p. 301-304. *In* T. Okaichi, D.M. Anderson and T. Nemoto (eds.) Red tides: biology, environmental science and toxicology. Elsevier Science Publishing, New York.

McMinn, A. 1987. Late Pleistocene dinoflagellate cysts from Bulahdelah, northern New South Wales. Proc. Linn. Soc. NSW. 109(3): 175-181.

McMinn, A. 1989. Late Pleistocene dinoflagellate cysts from Botany Bay, New South Wales, Australia. Micropaleontol. 35(1): 1-9.

Rossignol, M. 1964. Hystrichosphères du Quaternaire en Méditerranée orientale dans les sédiments Pléistocènes et les boues marins actuelles. Rev. Micropaleontol. 7(2): 83-99.

Sarjeant, W.A.S. 1982. Dinoflagellate cyst terminology: a discussion and proposals. Can. J. Bot. 60(6): 922-945.

Steidinger, K.A., L.S. Tester and F.J.R. Taylor. 1980. A redescription of *Pyrodinium bahamense* var. *compressa* (Böhm) stat. nov. from Pacific red tides. Phycologia 19(4): 329-334.

Wall, D. and B. Dale. 1968. Modern dinoflagellate cysts and evolution of the Peridiniales. Micropaleontol. 14(3): 265-304.

Methods for Dinoflagellate Cyst Studies

(Adapted from Matsuoka, Fukuyo and Anderson 1989)

Sediment Sampling for Dinoflagellate Cysts

The choice of sampling sites should be based on an inspection of local bathymetric maps and knowledge of local hydrography. Cysts are being accumulated, resuspended and transported in the same way as fine silt or mud particles. Accordingly, black undisturbed sediments from deep basins offer better opportunities for cyst studies than coarse sandy sediments characteristic of strong current regimes.

A variety of suitable gravity corers such as the Phleger bottom sampler and piston corers are commercially available but these instruments can also be easily home-built. Bottom samplers such as dredges or grab buckets which often lose the light fluffy material at the sediment surface (including many fresh living cysts) are less suitable. During the workshop SCUBA diving methods and a light-weight TFO gravity corer (Fig. 1) (Fukuyo and Matsuoka 1987) were demonstrated. Upon retrieval, the top and bottom of coring tubes must immediately be capped to prevent water leakage. Cores can then be stored in the cold (4°C) and dark until further analysis.

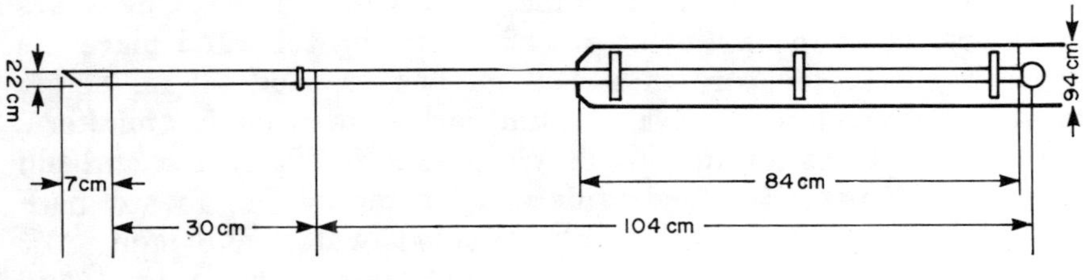

Fig. 1. Diagram of TFO gravity corer.

Sediment Processing

Two different methods can be used for cleaning and concentrating cysts from sediments - a standard palynological technique that uses harsh chemicals and a sieving technique that uses no chemicals.

Sieving Technique

For most red tide studies in which living cysts are required for germination, the sieving technique is the method of choice. Push the sediment out of the coring tube from below and suspend the upper 2 cm of the core into filtered seawater. Use an ultrasonic probe or bath to disaggregate sediments and cysts. Remover the larger particles by sieving through 250 µm and 125 µm sieves and collect cysts (20-80 µm size) on to a 20 µm sieve. Transfer the residue from the 20 µm sieve to a watch glass. Make a water-eddy in the watch-glass to separate heavier sand grains from cysts and other light-weight particles which will be concentrated at the centre (panning in the geological sense) (Fig. 2).

Palynological Technique

Where cysts are sparse (Matsuoka et al., this vol.) palynological processing techniques can produce more concentrated samples for study. This method uses dangerous acids and adequate safety precautions of working in a fume hood with rubber gloves are necessary. Push the sediment out of the coring tube and place the upper 2 cm of the core in a 100 ml beaker. Add 20 ml of 5% hydrochloric acid to remove calcium carbonate from foraminifera, shell fragments, corals and others. Wash with distilled water and add 10 ml of a 1% potassium hydroxide solution and heat in a water bath (70°C) for 3 mins. Wash with distilled water and add 30% hydrofluoric acid to remove silicate materials such as sand and diatom valves. Heat in a water bath at 70°C for 2 to 3 hrs in a fume hood. The residue should be neutralized with calcium carbonate. Wash again with distilled water and concentrate the material through a 125 µm sieve onto a 20 µm sieve. The cleaned cysts can then be mounted on a glass slide with glycerine jelly.

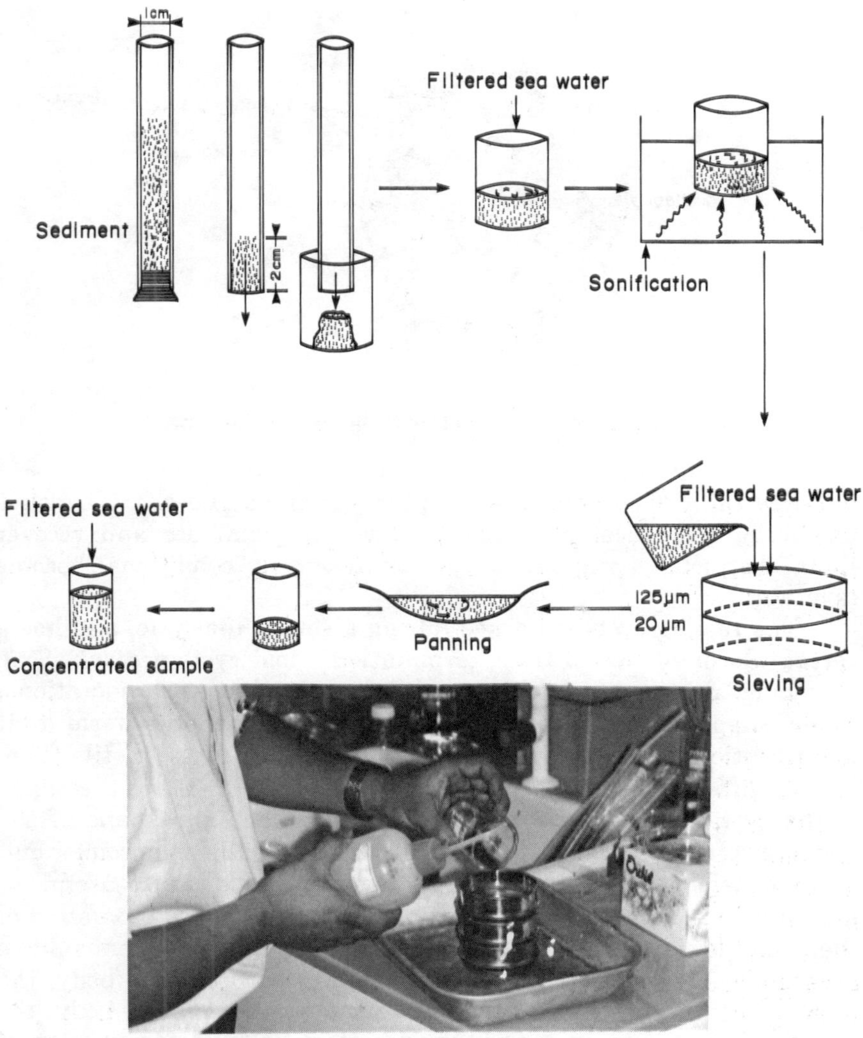

Fig. 2. Sediment processing for cyst study, showing a flow diagram (upper) and the method of washing the sediment through a series of 250 µm, 125 µm and 20 µm metal sieves to concentrate cysts.

Cyst Identification

Two types of cysts, temporary and resting cysts are found. The temporary cysts are formed under non-suitable environmental condition by cutting off the flagella from the motile cells. Sometimes they shed their theca and cell membrane to transform into round

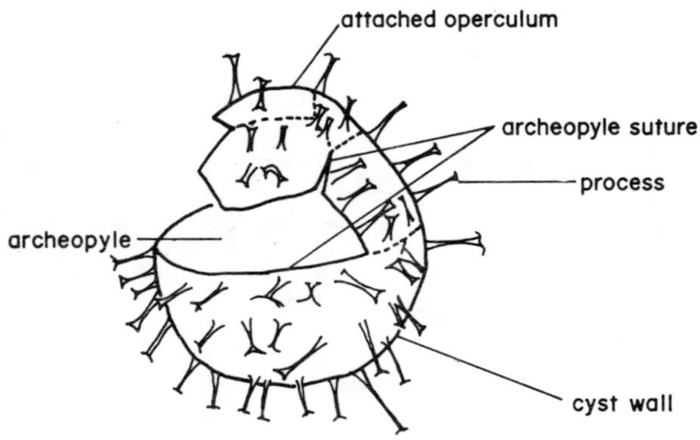

Fig. 3. Morphological features used for cyst identification.

ball-like cells. Therefore, the temporary cyst has no characteristics useful for identification. Usually they can germinate and recover their distinctive original shape as soon as conditions become favorable.

The resting cysts is formed during a sexual life cycle, and has a certain dormant period from germination. Most cysts are spherical, ellipsoidal or polygonal with or without spine-like ornamentations. Their shape is so different from that of the motile form that identification, i.e., determination of species name of the motile form, is very difficult. The important characteristics (Fig. 3) are the shape of the cyst, ornaments, wall structure and color, and shape of the opening, termed archeopyle, through which protoplasm comes out. Arrangement of ornaments such as lines of spines is also useful for identification, because the arrangement often expresses tabulation of their motile forms. As cysts belonging to the same genus have archeopyles of similar shape at the same position of the body, the archeopyles of similar shape at the same position of the body, the archeopyle is the most important characteristic for determination of genus. A useful guide to cyst identification has been prepared by Matsuoka et al. (1989).

References

Fukuyo, Y. and K. Matsuoka. 1987. *In* A guide for studies of red tide organims, Japan Fisheries Resource Conservation Association, Shuwa, Tokyo, p. 85-101.

Matsuoka, K., Y. Fukuyo and D.M. Anderson. 1989. Methods for modern dinoflagellate cyst studies, p. 461-479. *In* T. Okaichi, D.M. Anderson and T. Nemoto (eds.) Red tides: biology, environmental science and toxicology. Elsevier Science Publishing.

Sampling Coastal Dinoflagellate Blooms: Equipment, Strategies and Data Processing

PETER J.S. FRANKS
DONALD M. ANDERSON
Biology Department
Woods Hole Oceanographic Institution
Woods Hole, MA, USA 02543

FRANKS, P.J.S. and D.M. ANDERSON. 1989. Sampling coastal dinoflagellate blooms: equipment, strategies and data processing, p. 235-256. *In* G.M. Hallegraeff and J.L. Maclean (eds.) Biology, epidemiology and management of *Pyrodinium* red tides. ICLARM Conference Proceedings 21, 286 p. Fisheries Department, Ministry of Development, Brunei Darussalam, and International Center for Living Aquatic Resources Management, Manila, Philippines.

Abstract

Brief descriptions of a variety of low-cost systems for sampling dinoflagellate populations are given, together with sampling schemes for the field, and the factors which will affect the sampling program. Methods of data collection and reduction are discussed, and a brief introduction to methods of data interpretation is made.

Introduction

Dinoflagellates are important constituents of the coastal phytoplankton community throughout the world. Under conditions which are still poorly understood, certain species can dominate the phytoplankton community, creating a visible coloration of the water: a "red tide". The ability of these organisms to swim may contribute to the formation of such dense blooms by allowing the dinoflagellates access to deep reservoirs of nutrient-rich water. However, swimming creates particular problems in sampling dinoflagellates and in interpreting their vertical and horizontal distributions.

It is becoming increasingly apparent that much of the large-scale variability in dinoflagellate distributions can be explained by the advection and diffusion of blooms by physical processes. These processes include wind-driven, buoyancy-driven and tidally-generated motions of the water column, coupled with seasonal cycles of temperature and freshwater runoff. At a minimum, dense CTD (Conductivity, Temperature and Depth) coverage in time and space is necessary to understand the forcings and resultant motions. These samples should be coincident with biological sampling to provide maximum insight into the couplings of the physical and biological systems.

Many questions arise: what is sufficient coverage, what types of samples should be taken, and when should they be collected? Here we offer some insights based on our experience in sampling coastal dinoflagellate populations in the Gulf of Maine, USA. First we describe sampling methodologies: the choice and construction of sampling gear, the types of samples to take, and the interfacing of different types of instruments. We then describe sampling strategies: the selection of stations in both time and space, and data reduction and interpretation. We hope this information will be useful to those designing field programs in coastal areas, and that our guidelines will lead to sufficient and interpretable data sets that can be compared to similar data sets from other regions.

Sampling Methodologies

Dinoflagellate populations normally show a great deal of vertical structure in natural waters. In many situations this structure is intimately linked to the hydrography, the cells being found within the seasonal pycnocline or zone of maximum vertical density change. In other cases, the cells may accumulate quite independently of the hydrography due to their ability to swim. Sampling gear must be capable of sufficient vertical resolution to distinguish these possibilities. This often requires biological sample spacing of 5m or less in the vertical. Here we will describe a variety of sampling systems which are capable of dense vertical resolution for a range of field situations.

Pumping Systems

In shallow (<5m) embayments, a simple integrating tube sampler may be the most appropriate sampling device. These samplers consist of a length of 6.25-cm internal diameter (i.d.) PVC pipe, fitted with a cork attached to a line threaded through the pipe to a handle at the top (Fig. 1). A short foot at the bottom keeps the sampler from being pushed into the benthos. The sampler is slid vertically into the water until the foot rests on the bottom. Once the depth to the bottom is known, the tube is raised and moved to a new location nearby, and lowered until the foot is just off the bottom. This procedure prevents the water column samples from being contaminated with resuspended material. The line is then pulled to seal the tube with the cork. The tube is emptied by pouring from the top, since pouring from the bottom causes spillage around the cork. The integrated vertical sample can be poured into a carboy or bucket for further subsampling (see below). The advantages of this type of sampler are: 1) low cost; 2) ease of construction; 3) one sample integrates any vertical heterogeneity of the organism; and 4) ease of deployment. The disadvantages include: 1) no vertical resolution; 2) relatively small volume; and 3) no real-time vertical information.

In shallow areas, and relatively calm seas, the tube sampler described by Lindahl (unpub. ms.) may be the most appropriate sampling device. This low-cost sampler consists of lengths of 2-cm i.d. PVC pipe or garden hose, linked with valves and easily-separated connectors (Fig. 2). The length of a section determines the vertical

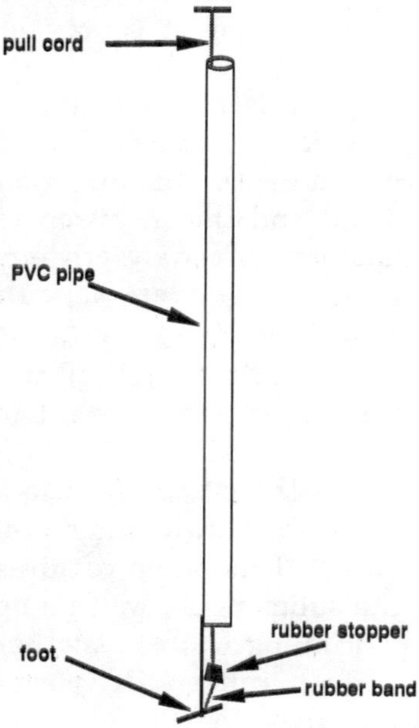

Fig. 1. The tube sampler. When the cord is pulled at the top, the cork seals the bottom of the tube. The rubber band keeps the cork oriented properly until the cord is pulled.

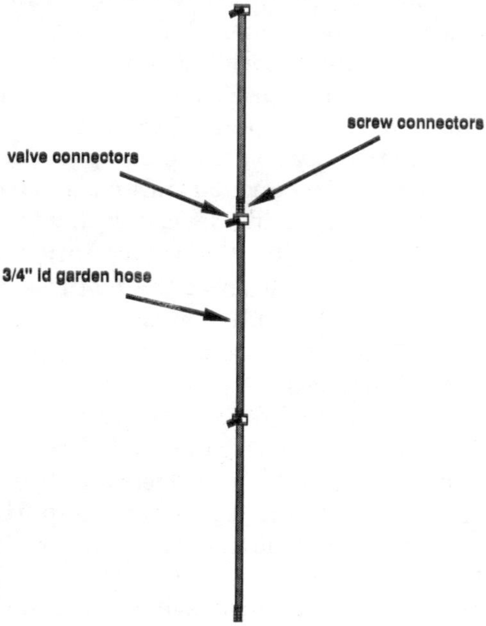

Fig. 2. The hose sampler. Lengths of hose are joined by valves and connectors. All valves must be open during deployment, and the surface valve closed upon recovery. A valve must be closed before the section of hose above it is removed, or the sample will be lost.

resolution of sampling. The sampler is slowly lowered with all valves open until the hose is filled. The top valve is then closed, and the hose raised until the next valve down can be closed. The upper section can now be removed, and its water drained. This procedure can be repeated until the whole length of hose has been raised. Hydrostatic forces within the hose will hold the water within the hose while it is being raised. Thus a small-volume vertical profile of the water column is obtained. The advantages of this sampler are similar to those of the tube sampler described above, although vertical resolution is obtained. The disadvantages are: 1) small volume; 2) no real-time vertical information; and 3) smearing of vertical structure within the pipe due to its narrow diameter.

Most vertical areas are dynamically complicated, requiring detailed coverage of both the biology and the hydrography. The samplers described above can be useful, but are nevertheless somewhat limiting due to their small volume and inability to provide continuous vertical profiles. We suggest the use of a pump profiling system. These systems have the advantages of: 1) high volume, permitting sampling of a number of variables; 2) continuous vertical profiles; and 3) the possibility of real-time observations of vertical features. The disadvantages are: 1) relatively high cost (generally < US$200); 2) they are more difficult to deploy than the sampler described above; and 3) they require an AC or DC power source for the pump and associated instruments.

The pump profiling system at its simplest is a pump and a length of hose. Aspects of the system design which must be evaluated include the position of the pump (at surface or at depth), the insertion of various flow regulators and a bubble trap, and the inclusion of subsidiary sampling devices such as fluorometers, autoanalyzers, etc. Here we describe the configuration used in our laboratory, with justification for each of the features. This configuration is certainly not exclusive of other arrangements; we hope that readers will use the information here in designing pump profiling systems specific to their own needs.

The central feature of the pump profiler is the pump. We use a "Lil' Giant" submersible pool pump (available from swimming pool supply companies, and scientific supply houses). This pump requires AC power, as would another alternative, the submersible well pump (available from dealers who drill wells for drinking water). Another feasible pump is a boat bilge pump which can run on DC power (available from marinas and boat supply companies).

The pump need not be submersible, but it should be able to withstand being soaked in salt water. If a deck pump is used, it should be self priming: priming the pump is the most difficult aspect of the deployment of the Lil' Giant system. The pump can be suspended either just below the surface or at the bottom of the hose. The consideration here is the formation of bubbles due to cavitation within the pump. The surface pump is likely to cavitate if the hose below is too long. We use 40 m of 2-cm i.d. garden hose, the maximum length useable with this pump. Placing the pump at the base of the hose solves the cavitation problem, but generates a new problem with the length of electrical wire over the side of the vessel. Kinks in the hose and the tangling of wires are the most serious deployment problems after the flow has started.

Since the Lil' Giant pump is not self-priming, the hose must be completely filled with water before the pump is plugged in. After much trial and error, we found that putting the hose over the side of the vessel until it is fully submerged is the best way to fill it: this means that all our vertical profiles are taken as the hose is raised. The hose must be deployed carefully, as it has a tendency to twist and kink. Kinks are fatal to a profile. The most common location of kinking is where the hose attaches to the wire. To ease the strain here, we use a dual thickness of radiator hose, clamped to the wire with hose clamps, with a screw connector for the garden hose at the other end (Fig. 3). This is a particularly convenient arrangement, since the hose may be unscrewed at any time if the wire is required

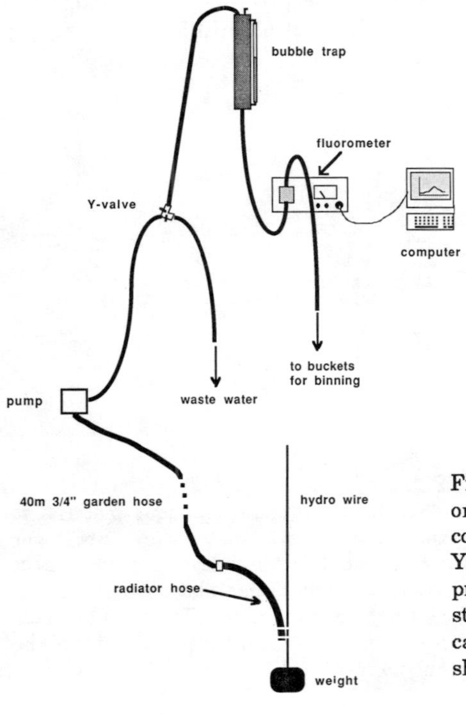

Fig. 3. The pump profiling system. This is only one of a variety of possible configurations. The flow is regulated at the Y-valve. The radiator hose at the wire prevents the hose from kinking and stopping the flow. A variety of instruments can be included in-line; a fluorometer is shown here as an example.

for something else, without having to remove the radiator hose and hose clamps. Any reasonably stiff length of hose can be substituted for radiator hose, so long as it will not kink. It is also useful to have duct tape handy: it can be used as a hose wrap should the hose kink or fail. Note also that a heavy weight is required to keep the hose as vertical as possible. We use a 30 kg weight attached to the wire (never to the hose itself).

The hose leading out of the pump brings the water on deck. A Y-valve is necessary for dividing the flow into the portion to be sampled, and the waste, which is pumped overboard. This Y-valve is critical to maintaining a steady flow to the instruments. Steady flow is achieved by adjusting the amount of waste water, not by adjusting the flow of water to the instruments. The valve to the instruments should always be wide open to avoid bubble formation.

Bubbles within the hose can be a serious problem with certain instruments such as fluorometers and autoanalyzers. To mitigate the problem, we include a bubble trap between the Y-valve and the instruments (Fig. 4). The bubble trap consists of a 1-m length of 10-cm i.d. acrylic pipe, fitted with stoppers and hose connectors at the top and bottom. A small chimney tube at the top allows air to escape. A length of clear tubing joined to the top and bottom with right-angled connectors allows visualization of the water level within the pipe. A 20-cm length of hose extends from the top hose connector into the bubble trap. The level of water within the bubble trap should

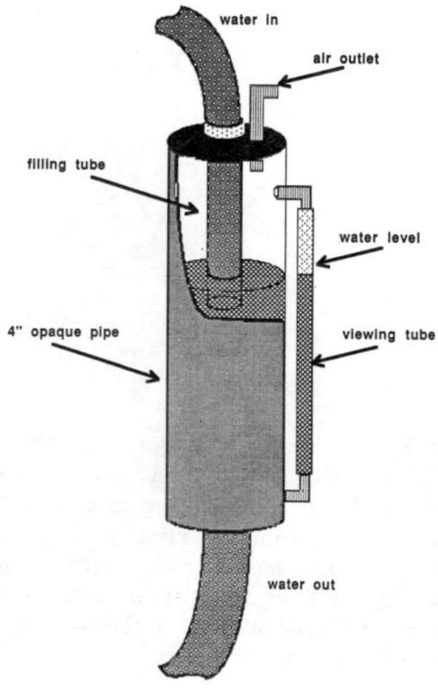

Fig. 4. Cut-away view of the bubble trap. This should be made of opaque material so that the phytoplankton are not over irradiated. The water level should be kept above the end of the inlet hose so that no new bubbles are created. The level of water is shown by the clear tube on the side of the bubble trap.

always be kept above the level of this hose so that no more bubbles are created. The bubble trap itself should be located at the highest point of the pumping system, so that all bubbles within the system may escape. The height of the bubble trap will largely be determined by the head of the pump and the geometry of the ship.

From the bubble trap, the flow feeds by gravity into the attached instruments, and back to the deck. The water may be collected in carboys to obtain samples integrated over any desired depth interval as the hose is raised or lowered. For certain analyses (e.g., productivity or chlorophyll) the sample containers should be opaque or acid-washed, although a seawater rinse is sufficient for species counts.

The protocol for sampling with the pump profiling system is complicated by the residence time of the water within the hose. Thus the water which is being pumped on deck was obtained from the depth the hose inlet occupied some time ago. The method we use to correct for this requires measurement of the residence time. This was done using a flow-through fluorometer to detect a spike of chlorophyll introduced at the hose inlet. Simple coloured dye would work, as long as it is visible to the eye or to a spectrophotometer after passage through the hose. The calculated transit time should be at the first appearance of the dye at the end of the hose: smearing within the hose will cause an initial spike to be spread over a considerable distance within the hose.

Once the transit time within the hose is known, the procedure for performing a vertical profile is relatively easy. The variables which must be decided before the profile is begun (and preferably before the cruise begins) are: 1) the rate of rise of the hose; and 2) the flow rate. The former will depend mostly on the winch being used. We have found that a rate of 2 m/min. gives reasonably dense vertical coverage. However, most winches will not raise that slowly. Our protocol is thus to raise the hose 1m every 30 seconds, and pause at that depth. The hose inlet usually takes 5 or 6 seconds to rise 1 m. Thus the profile is taken in a series of steps. Water from 5-m intervals is collected in individual buckets for subsampling. Clearly, it is important to time the raising of the hose and the bucket collection carefully so that a representative sample is collected that integrates over the 5-m interval.

In order to know the depth from which the water that arrives on deck was pumped, the winch operator and the person handling the hose must work independently. At the start of the profile, both persons should start their stop watches. The winch operator keeps his going, raising the hose 1 m every 30 seconds, beginning 30 seconds after the start signal. The person handling the hose waits an amount of time equal to the transit time of the water in the hose, at which time the water being pumped on deck is water for the first interval. At that time, the hose handler should restart his stopwatch, to time the filling of the bucket. For example, to integrate over 5 m, the hose handler would place the hose outlet into a bucket for 2.5 min. (if the hose is being raised at 2 m/min.). Continuing this process will give samples binned into 5-m intervals. The winch operator must pause at the surface so that the hose handler can "catch up". The hose inlet must not break the surface, or the pump's prime will be lost.

Various strategies for binning or integrating of samples will be described below. Subsampling from the bins depends on the information needed. We sieve one liter from each bin through 20-mm Nitex mesh (epoxied onto a cylinder of 8-cm PVC pipe) and preserve it in 5% formalin for cell counts. An additional 500 l is filtered through GF/A filters. The filters are used for chlorophyll analyses and the filtrate is frozen for nutrient analyses. The flow rate should be adjusted at the Y-valve so that sufficient water is obtained for all these analyses, as well as for the auxiliary, on-line instruments.

Auxiliary Instruments

A variety of auxiliary instrumentation is available for interfacing with the pump profiling system, depending on the needs of the project and the available budget. Many of the instruments are relatively costly (> US$5000), and so may not be available to smaller laboratories.

Our particular pumping system includes an in-line fluorometer (Turner Designs Model 10-00R with flow-through cell and chlorophyll *a* filter set). This fluorometer is interfaced to a portable personal computer (NEC APC IV Powermate Portable) through an A/D (analog/digital) board (Metrabyte Dash-8; Metrabyte Corp., 440 Myles Standish Blvd., Taunton MA, 02780 USA) following the instructions in the fluorometer manual. The excellent software supplied with the A/D board was modified to plot fluorescence on the computer screen in real time, and to store the data to disk. The screen plot allows visual location of the fluorescence maximum, and can be used in real time to modify sampling and "binning" procedures. We have found the vertical fluorescence profiles to be an invaluable tool in our phytoplankton field studies, even though the dinoflagellates of interest are seldom the dominant source of fluorescence.

Alternative in-line devices could include an autoanalyzer (e.g., Technicon AutoAnalyzer II, available through scientific supply companies) to obtain detailed nutrient profiles. A transmissometer could be included in-line to obtain turbidity data. Some sort of flow chamber must be designed for the transmissometer which is especially sensitive to bubbles. We suggest using a submersible transmissometer, located below the hose inlet.

The A/D board may have additional available channels. These can be used to digitize the signal from a light meter or pressure sensor attached near the hose inlet. Once again, these instruments require additional wires, which can cause tangling and kinking of the hose.

As mentioned above, we consider it essential to obtain coincident hydrographic and biological information. It is somewhat unfortunate that most dinoflagellate blooms are associated with strong signals in salinity: salinity is much harder to measure than temperature. Temperature is sometimes a useable surrogate for salinity, but not reliably. In dynamically simple areas such as shallow embayments, salinity can be measured adequately with a hand-held, relatively inexpensive refractometer. This device has an accuracy of about 0.2 ppt, which is reasonable for most coastal situations. If no more sophisticated instruments are available, temperature readings, measured with a thermometer, and salinity values, measured with a refractometer, should be taken at least at 0.5 m intervals. These measurements can be made on water exiting the hose.

Inexpensive battery-operated temperature/conductivity probes, such as those manufactured by InterOcean Systems Inc. (3540 Aero Ct., San Diego, CA 92123 USA) can be lowered to learn details of water column structure, but accuracy is limited. A more sophisticated system might include an in-line thermosalinograph (available through InterOcean Systems, Inc.). This instrument may also be interfaced to the computer, allowing rapid data aquisition and storage. The quality and durability of such instruments is of paramount importance: nothing is more frustrating than trying to collect data with unreliable instruments. Always try to test an instrument in the field before purchase, and talk to others who have used the instrument.

The instrument we recommend is a CTD (Conductivity, Temperature, Depth) profiler. We use the "Sea Cat Profiler" (Sea Bird Electronics, 1808-136th Pl. NE, Bellevue WA, 98005 USA). This small CTD stores data internally during a cast, thus no extra electrical wires are required over the side of the vessel. The data can be subsequently transferred to a personal computer using the various programs supplied with the CTD. The instrument itself is practically indestructible and foolproof. We generally mount our CTD below the hose inlet, with the sensors pointing upward in order to take data as the instrument is raised through the water.

Station Selection: Spatial

The decision of when and where to locate stations can be overwhelming. Coastal areas have notoriously complicated physical systems, requiring relatively dense spatial coverage. The main forcing over much of the year is the wind, which varies on time scales of hours and days. What is the best strategy for dealing with these problems, without spending your whole life at sea?

The first suggestion we make is to plan as dense CTD coverage as possible. Hydrographic data are an absolute necessity for interpretation of any data concerning biological distributions in the ocean. One criterion for deciding on CTD station spacing is the internal Rossby radius of deformation, Ri. This length scale is the natural length scale for most physical features in the ocean: the width of frontal zones, the size of eddies, the width of river plumes, etc. (Franks, 1990, in press). It is calculated from the thickness of the surface layer, h, the densities of the upper and lower layers, r and r', the acceleration due to gravity, g, and the Coriolis frequency, f:

$$R_i = \frac{gh\,(\rho' - \rho)}{f}$$

This length scale is typically < 5 km in most coastal regions. Hydrographic station spacing of this order or less will allow resolution of features such as fronts and river plumes. Since most dinoflagellate blooms are associated with such features, it is important to be able to resolve them with some confidence.

One of the main factors contributing to the complicated dynamics of coastal regions is the physical barrier formed by the coast. This feature will tend to align physical systems (e.g., wind-driven upwelling, coastal currents and river plumes) parallel to the coast. This means that hydrographic variables will tend to show greater changes across shore than along shore. For this reason, it is important to obtain good cross-shore coverage. We suggest sampling to at least 5Ri from the coast, in order that coastal features be adequately sampled.

Most physical systems in coastal areas tend to follow bathymetric contours. At areas where the curvature of the coast is very sharp, a physical feature such as a coastal current will separate from the coast and move into open water. The criterion for this separation is roughly given by the Rossby number, R_o:

One of the main reasons for recommending a CTD profiler is that it may be used without the hose pumping system. A vertical profile with the pumping system may take half an hour, whereas a CTD cast need only take a few minutes. Thus dense CTD coverage may be obtained in an area with more sparse biological profiles. The variations in hydrography can be used to explain details of the biological distributions and dense CTD coverage will always make data interpretation easier. An additional advantage of many CTD profilers is that they may be expanded to include *in situ* fluorometers, transmissometers, light meters, O2 sensors, etc. These instrument packages are easily deployed even in fairly rough seas, when deployment of a pumping system may be impossible. They also allow dense sampling of a variety of fields.

One final word on sampling equipment: always plan for the worst. Bring spares of all pieces of equipment: hose, pumps, valves, connectors, radiator hose, hose clamps, etc. A supply of duct tape is a necessity. Always make contingency plans if any aspect of the sampling program should fail at sea, e.g., if the pump loses its prime, the computer fails, or the fluorometer breaks. Planning ahead for such emergencies will help to make the best out of a bad situation, and may prevent the waste of a lot of time and money.

Sampling Strategies

Here we describe various factors to consider when designing a sampling program. We examine certain physical parameters which merit consideration, and suggest some biological features which may influence timing and location of stations. Finally we describe some standard techniques of data reduction and interpretation.

$$R_o = \frac{u}{fr_c}$$

Here u is the velocity of the current, f is the Coriolis frequency for that latitude, and rc is the radius of curvature of the coast (or the bathymetry). If this number can be calculated, and it is greater than 1, the coastal current is likely to separate from the coast and move offshore. The sampling scheme should be adjusted accordingly, and more offshore samples taken. The orientation of the transects may also be adjusted: they should, in general, be oriented perpendicular to the feature being sampled. Near a sharp bend in the coast, the transects may be oriented almost parallel to the coast in order to obtain good hydrographic coverage of a coastal current (see Fig. 5).

If multiple transects are to be run, it is preferable that they be oriented parallel to each other. This will give even data coverage,

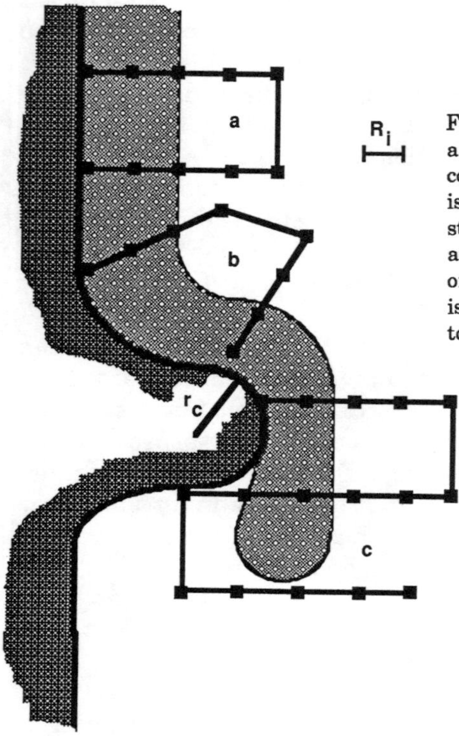

Fig. 5. An example of a sampling scheme for a coastal current which separates from the coast at a cape. In section a, where the coast is straight, the legs are parallel, and the stations evenly spaced. In section b, the legs are oriented perpendicular to the coastline or the bathymetry. More offshore coverage is made in areas where the feature is likely to separate from the coast such as section c.

with no poorly sampled areas. It may be more efficient, in terms of the ship, to orient the transects into a Z shape. However, relatively large gaps in coverage occur at the top and bottom of the Z. A more even coverage is obtained with an E-shaped cruise pattern. This will be found to be important when trying to contour the data and make surface maps of features.

In the vertical, dense coverage is always preferable. Some workers advocate binning of samples into three categories: below the thermocline, within the thermocline, and above the thermocline. We do not suggest using this procedure for several reasons: 1) the thermocline may not correlate with the dinoflagellate peak or the nutricline; 2) the sample resolution is low; 3) the variable depth of the thermocline will make data reduction and interpretation difficult; and 4) unexpected features may be missed. Rather, we recommend obtaining as many evenly-spaced samples as is feasible given sample processing time. The even spacing allows for quick and easy plotting of data, with good resolution of most vertical features. We use a binning interval of 5 m, and recognize that this spreads the cell concentrations and nutrients out vertically. Thus, a feature which is

1-m thick in the *in situ* fluorescence becomes 5-m thick in the cell counts.

We also stress obtaining coincident CTD data. These data will help to resolve binning problems by giving a detailed picture of the vertical structure of the water column. The more different types of data obtained in vertical profiles, the easier will be the interpretation of the data. For example, we have found transmittance to be a good inverse tracer for *in situ* fluorescence. Since they are obtained by very different methods (transmittance via submerged instrument, fluorescence via the hose), we are confident that the strong correlation between the fields is real. Thus the transmittance gives an independent check on the hose profiling system.

Station Selection: Temporal

Considerations for the timing of sampling are numerous. Physical factors such as wind, rain, tides and sunlight have all been shown to affect dinoflagellate distributions. Seasonal considerations are also important, as dinoflagellates often show fairly restricted periods of extensive growth.

The linkage of dinoflagellate distributions with hydrographic features is strong in both space and time. Forcing such as the wind and tides will have effects on both the hydrography and the dinoflagellate distributions and should be taken into account when planning cruises. However, for most sampling programs the size of the ship is the main consideration in timing cruises: a small vessel cannot be used in heavy seas and deployment of a pump profiling system may be impossible in even moderate swell. This causes aliasing with respect to the wind: cruises are only made in light seas. Our solution to this has been to make multiple cruises, as often as possible, and to interpret the CTD data with reference to continuous wind data obtained nearby.

Tides are more predictable, the semi-diurnal and fortnightly tides being the most prominent. A good knowledge of the local tides is important in deciding whether tidal advection or mixing will be a serious problem. In small embayments, the tides may be the predominant mode of forcing, while in shallow coastal areas, strong tides may create fronts which accumulate phytoplankon. If possible, samples should be taken at the same phase of the tide each time. For large-scale features, the tides may be relatively unimportant. At the very least, the phase of the semi-diurnal tide should be recorded at each station, in order that any aliasing be taken into account during data reduction.

Tides are also important when sampling continuously at a single station, for example when performing vertical migration studies. For this reason, samples should be taken at an interval of 3 hours or less, in order that all phases of the semi-diurnal tide be resolved. This may lead to a large number of samples to process, but the data will be less subject to misinterpretation.

Over a season the frequency of cruises depends largely on the organism being sampled and its suspected distribution. In general, though, at least one cruise should be taken before the particular dinoflagellate blooms, in order that the initial conditions of the water masses are known. When studying the distribution of a particular dinoflagellate, it is best to concentrate cruises early in the bloom. This will allow good resolution of the physical systems mediating bloom distribution as the bloom develops. In particular, numerous cruises can help distinguish between in situ growth, and alongshore advection, a particularly difficult problem in much dinoflagellate research.

If the dinoflagellate being studied is toxic, the local shellfish or fish monitoring programs can provide valuable information for planning the location and timing of cruises. We rely heavily on the state-run shellfish toxicity monitoring program for planning cruises: the timing of toxicity indicates how often cruises should be taken, while the spread of toxicity determines the extent and location of a given cruise. As the bloom spreads, we alternate between cruises along one transect, gathering detailed vertical information, with cruises having extensive alongshore coverage (in the E pattern), but less vertical resolution. In the latter instance, we are able to sample 25 stations (continuous CTD profiles, 2 bottle casts for cell counts at each station) from a fast (18 kt) 10-m vessel in 8 hours. Our single transect with more detailed data at each station covers only 5 stations in the same time from a 15-m ship.

In the absence of a monitoring program, there are some environmental cues which correlate well with dinoflagellate blooms. Most dinoflagellate blooms are found in the pycnocline of a well-stratified water column. This stratification can be caused either by salinity differences or by heating. Thus strong rains or several sunny days in a row can be important in bloom formation. Keeping a close eye on the weather is important when sampling for dinoflagellates.

Data Reduction: Smoothing, Plotting and Interpretation

The detailed vertical coverage obtained by a pump profiling system with a CTD is both a boon and a bane: the large amount of data provides an excellent picture of the vertical distributions, but data processing can be tedious and time-consuming.

One of the main problems in using a variety of instruments is the diverse nature of the data sets generated. Merging these data sets into a single visualizable data set can be a difficult task. The techniques of data smoothing and interpolation become indispensable at this stage. Data smoothing is generally necessary to remove instrument spikes and noise. A variety of tools exists for smoothing data; we recommend the use of cubic splines. An excellent book, "Numerical Recipes" (Press et al. 1986) describes the use of a variety of splines, and gives computer codes for their implementation. A useful offshoot of the smoothing process is that the smoothed data are easily interpolated. Thus vertical profiles obtained with a variety of instruments can be interpolated onto the same vertical grid, and merged into a single data file.

Data gathered via the pumping system (e.g., fluorescence) will generally be stored as a time series, whereas CTD data are plotted versus depth. To merge a time series with a depth profile, some simple manipulations of both data sets may be necessary. First, the CTD data and the time series data must be converted to the same sampling rate. Thus, if the CTD samples at 0.5-second intervals but the fluorometer samples at 2-second intervals, every four CTD data points should be averaged. The unwanted portions of each data set should then be removed (i.e., the CTD up- or downcast data and the fluorescence data up to one hose-transit-time from the start of the profile). The two data sets can then be merged, line for line, creating a single data set. This data set can then be smoothed and interpolated onto a vertical grid for plotting depth profiles.

Discrete data such as cell counts, chlorophyll samples, or nutrient data can be smoothed and interpolated in the same manner as the CTD data. However, the low vertical resolution makes this an unreliable method of data manipulation. We prefer plotting such data as histograms in a vertical profile (Fig. 6). This method leaves no ambiguity as to sample location and lets the readers draw their own conclusions as to adequacy of sampling. Fig. 6 demonstrates another useful property of cubic splines: they can be used for calculating gradients of quantities. Once the coefficients for the spline are known, it is easy to differentiate them to find the slope of a property. This process is described in Press et al. (1986) and computer codes are given for implementation. In the case of Fig. 6, we have calculated the Brünt-Väisälä, frequency, or vertical gradient of density, N^2:

$$N^2 = \frac{g}{\rho} \quad \frac{r\,\partial\rho(z)}{\partial z}$$

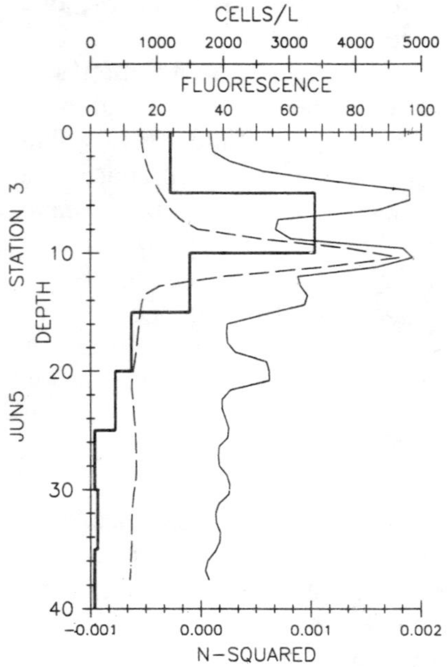

Fig. 6. Vertical profiles of cell concentration (*Ceratium longipes*, cells/l, heavy solid line), *in situ* fluorescence (dashed line), and N^2 (light solid line). As described in the text, maxima in N^2 correspond to maximum gradients in density (pycnoclines). The cell counts, binned into 5-m intervals, show a maximum coincident with the peak of *in situ* fluorescence, and the deeper pycnocline.

Here g is the acceleration due to gravity, r(z) the vertical density profile, and z the vertical coordinate. This quantity will show a maximum where the density gradient is strongest (i.e., the pycnocline). Thus it is very useful for assessing correlations between the density and other fields. As can be seen in Fig. 6, the maximum cell concentrations are found at the maximum gradient in density.

Plotting data is an art in itself. Numerous plotting packages exist for personal computers which allow quick and easy visualization of data. However, many of these packages will not allow plotting in the standard oceanographic format: a decreasing vertical axis. One way around this is to manipulate the data so that the depth is negative, with the surface being zero. This is not a very elegant solution, but cannot be avoided in many cases.

Plotting the vertical profiles of several properties on a single set of axes is a very useful means of visualizing the vertical correlations of various fields. Such a plot is shown in Fig. 7, where the raw and smoothed data sets are shown for comparison. The strong correlation between the phytoplankton fluorescence and the water density becomes obvious. This plot also shows some of the problems associated with smoothing data. These data were smoothed using an objective mapping routine after Levy and Brown (1986). The main problems associated with smoothing appear at the boundaries of the data set, in this case the surface and the bottom. In Fig. 7 it can be

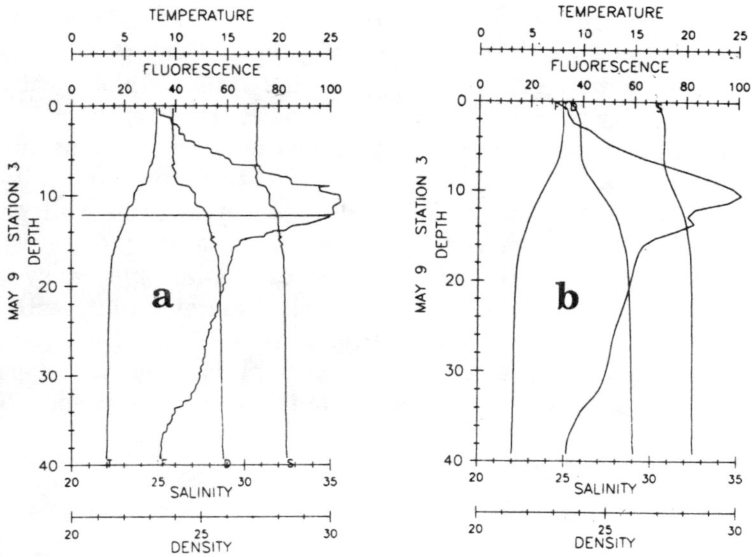

Fig. 7. Plots of the vertical profiles of temperature (°C), salinity (ppt), density (σT), and in situ fluorescence (relative units). Panel a shows the raw data (approximately 500 points in each profile), while panel b shows the smoothed profiles (50 points per curve). The small letter at the top or bottom of the profile indicates the property being plotted (T,S,D or F). Note that the smoothed salinity and density profiles tail off at the surface, while the raw data are vertical. This is an artifact of the smoothing program used.

seen that the smoothed data set shows a gradient at the surface, whereas the raw data show no such slope. This occurs because of the low number of data points near boundaries. Cubic splines can overcome some of these problems, but it is important to check the smoothed data against the raw data before conclusions are drawn. The second point where the smoothing program fails is in removing the instrument spike in the fluorescence at 12 m. Since we know that this spike was due to the instrument changing scales, and was not a real feature, we could manually remove the spike from the raw data before smoothing.

If many vertical profiles have been generated, it may prove fruitful to examine horizontal variations in properties. In this case we recommend contouring of data. Contouring of vertical profiles is a surprisingly difficult and frustrating task. If the stations were not evenly spaced, some interpolation scheme must be available for mapping onto a regular grid. If the depths of profiles vary because of bathymetry, some method should be available for masking out the bottom. Without this, the contouring program will generally create its own data in this region.

Given these caveats, contouring is still an extremely powerful tool in oceanographic research. A series of contour plots of some of our own data are shown in Fig. 8. The strong correlations of the diverse fields are immediately apparent from this type of plot, and the vertical variability is easily visualized.

The highlighting of vertical variations between stations is the most important aspect of contour plots. Most of this variability is caused by physical forcings; different types of forcings will show characteristic patterns in the cross-shore hydrography (Franks, in press). With experience, recognition of the particular physical system becomes more routine. Fig. 8 shows a river plume front, with an associated population of *Alexandrium tamarense*. The salinity gradient indicates the presence of the river plume, and the slope of the pycnocline suggests that it is moving towards the reader. The

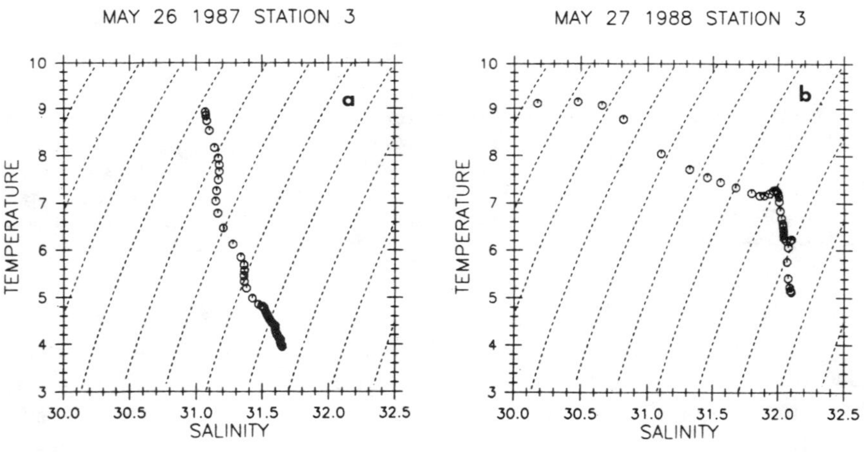

Fig. 8. Temperature/salinity plots from the same location, one year apart. The dashed lines are curves of constant density. Panel a shows a profile with a thermocline, while panel b shows a profile with a halocline. Note that the temperature range in the two panels is the same, but the salinity range is much less in a than b. The profile in b covers a much wider range of density.

magnitude of the slope allows an estimate of the current speed: ~10 cm/second. An introductory physical oceanography text will explain how these calculations are made.

Much information about dinoflagellate populations can be obtained by identification of water masses. This is most easily done using temperature/salinity (T/S) plots. The convention in oceanography is to make temperature the vertical axis, and salinity the horizontal axis. Plotting a vertical profile on such axes makes individual water masses readily apparent: deep water masses are cold and salty, thermoclines and haloclines will point at right angles to each other, and the surface water will be warm, fresh or both. Thus water below the pycnocline will cluster toward the lower right of the plot, while surface water will be toward the top or the left, depending on whether the pycnocline is due to temperature or salinity, respectively. In Fig. 9, two T/S plots are shown from the same location a year apart. Fig. 9a shows a typical profile with a thermocline, while Fig. 9b shows a profile with a halocline. The differences between these are obvious and demonstrate the variability possible in the hydrography of coastal regions. Such plots

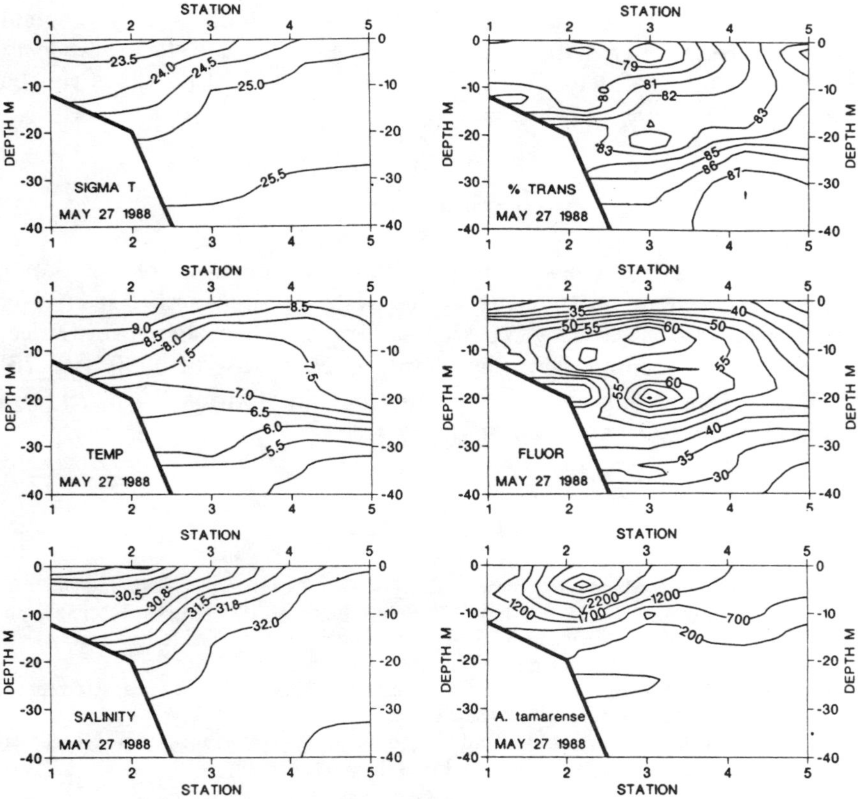

Fig. 9. A series of contour plots of data gathered in the Gulf of Maine. Clockwise from the upper left: density (σT), % transmittance, in situ fluorescence, cell concentration (Alexandrium tamarense, cells/l), salinity (ppt), and temperature (°C). The thick line at the bottom left of each panel is the bottom. The maximum depth is 40 m, the length of our hose. The stations are ~7 km apart; station 5 is 30 km offshore. Note the strong, sloping pycnocline created by the salinity gradient. The cells are found mainly within the fresher water and the pycnocline.

also bear strongly on the distributions of cells: in Fig. 9a, the pycnocline is formed by heating, so we might expect the cells to be horizontally uniform. In Fig. 9b the pycnocline is formed by a salinity gradient, indicating the presence of a river plume. We might expect considerable horizontal variability in cell concentrations due to their association with the river plume front.

Numerous other tools exist for visualizing data. Plotting a profile versus density rather than depth can be a useful way to assess correlations with the hydrography. Three-dimensional plots of density surfaces can lead to insights into the local dynamics. An introductory physical oceanography text, taking physical oceanography courses, or obtaining the help of a local expert for interpreting complicated hydrographic data are useful steps.

Of all the points made above, none can be stressed more than the importance of obtaining coincident hydrographic and biological samples. It is only with a good knowledge of the local physical dynamics that confident interpretations of biological patterns can be made.

Acknowledgements

Development of these methods and the preparation of this paper were supported in part by the Office of Naval Research (Grant No. N00014-89-3-1111) and by NOAA Sea Grant College Program Office, Department of Commerce under grant no. NA86AA-D-S6090 (R/B-76) to Woods Hole Oceanographic Institution. Contribution No. 7172 from the Woods Hole Oceanographic Institution.

References

Franks, P.J.S. Dinoflagellate blooms at fronts: Patterns, scales and forcing mechanisms. *In* L. Ford (ed.) Proceedings of the Fourth International Conference on Modern and Fossil Dinoflagellates, L. Ford (ed.) Amer. Assoc. Stratigraphic Palynologists Contr. Ser. (In press)

Levy, G. and R.A. Brown. 1986. A simple, objective analysis scheme for scatterometer data. J. Geophys. Res. 91:5135-5158.

Press, W.H., B.P. Flannery, S.A. Teukolsky and W.T. Vetterling. 1986. Numerical Recipes. Cambridge University Press, New York. 818 p.

1

2

3

4

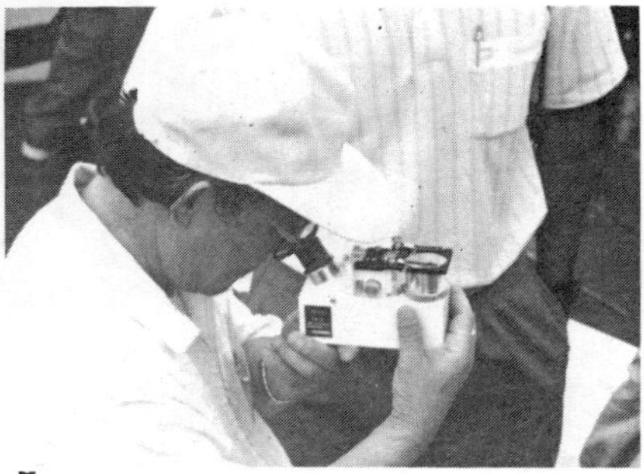

5

During the Brunei Darussalam workshop, a field trip enabled participants to see some field equipment in action. Photos show: 1. F.J.R. (Max) Taylor demonstrating his compact 20 µm plankton net. 2. Yasukatsu Oshima has a similar net. 3. Sherwood Hall uses a 5-m long net to harvest cells for toxin analysis. 4. Fernando Rosales-Loessener inspects a plankton sample in a capillary tube using a hand lens. 5. Tomotoshi Okaichi demonstrates a hand microscope.

6

7

8

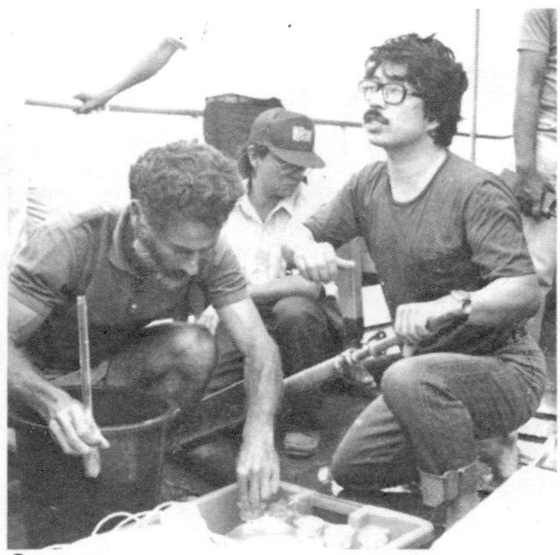

9

Using a sediment corer during the field trip. 6. Kelly Riroro about to release the corer, watched by Ranjith De Silva (left). 7. A participant from Brunei Darussalam extracts the tube containing the sediment sample, under the watchful eye of Yasuwo Fukuyo (right). 8. Webber Booth holds a core sample inspected by Ranjith De Silva (left) and Willy Pastor (right). Yasuwo Fukuyo is reassembling the corer. 9. Webber Booth and Yasuwo Fukuyo storing and labelling sediment samples.

Review of Culture Methods for *Pyrodinium bahamense*

SUSAN I. BLACKBURN
CSIRO Marine Laboratories
GPO Box 1538
Hobart, Tasmania 7001
Australia

YASUKATSU OSHIMA
Faculty of Agriculture
Tohoku University
1-1 Tsutsumidori-Amamiyamachi
Sendai 981, Japan

BLACKBURN, S.I. and Y. OSHIMA. 1989. Review of culture methods for *Pyrodinium bahamense,* p. 257-266. *In* G.M. Hallegraeff and J.L. Maclean (eds.) Biology, epidemiology and management of *Pyrodinium* red tides. ICLARM Conference Proceedings 21, 286 p. Fisheries Department, Ministry of Development, Brunei Darussalam, and International Center for Living Aquatic Resources Management, Manila, Philippines.

Abstract

The dinoflagellate *Pyrodinium bahamense,* both var. *bahamense* and var. *compressum,* has been successfully cultured in seawater-based media by several workers. These cultures were sometimes kept for many years but all of them were eventually lost. This tropical dinoflagellate requires temperatures of 24-30°C, dilute nutrient media and soil extract, while a pH optimum of 8 has also been observed. Recommendations for renewed culture attempts are provided.

Introduction

The establishment and maintenance of unialgal cultures is essential for studies of the causative mechanisms of toxic dinoflagellate blooms. Culture experiments with *Pyrodinium bahamense* are the only means to elucidate the growth characteristics and toxin production by this species under different environmental conditions and to describe the asexual and sexual life cycle including conditions for resting cyst formation and germination.

In the past thirty years, *P. bahamense* has been cultured at least four times (Table 1), in some cases for many years. The var. *bahamense* from the Caribbean has been cultured three times while the var. *compressum* from the Indo-West Pacific has been cultured only once.

A review of the culture methods used for *P. bahamense* is presented here, and recommendations for future culturing attempts are made.

Establishment of Cells in Culture

The establishment of dinoflagellate cultures is greatly facilitated by isolating cells during optimal growth conditions such as from blooms or red tides. The transport of viable *P. bahamense* cells from the sampling site to the laboratory is more likely to be successful if the proportion of *P. bahamense* cells to other phytoplankton is large. Samples should be diluted to avoid mortality from anoxia. Oshima et al. (1985) isolated cells from a 48-hour-old net sample which had been handcarried from Palau to Japan, diluted to 10-50 cells/ml with seawater from the collection locality. Finding sufficient cells to isolate is not a problem if the sample is from a bloom. McLaughlin and Zahl (1961) isolated *P. bahamense* from a bloom in Phosphorescent Bay, Puerto Rico. They micropipetted 8,000 cells into petri dishes, washed them in sterile seawater and eventually isolated groups of up to 100 cells.

Cell division may be promoted by using medium made up with seawater from the collection locality. This method was employed by Oshima et al. (1985) in isolating var. *compressum* from Palau, and by McLaughlin and Zahl (1961) for var. *bahamense* from Phosphorescent Bay.

Isolation success was found to be density dependent for cultures of var. *bahamense* from Oyster Bay, Jamaica (Seliger, pers. comm.). Best results were obtained by drawing single cells into a glass capillary tube which was then immersed in medium in a petri dish. The microenvironment thus created was more successful in promoting cell division than when using the petri dish alone.

Maintenance of Cultures

Culture media. The different media which have been used in culturing *P. bahamense* are summarized in Tables 1 and 2. Soil-based

Table 1. Summary of culture attempts of *Pyrodinium bahamense*.

Variety	Source material	Date	Medium	Temperature	Light intensity	Photo-period	Transfer time	Cultures held	References
var. *bahamense*	Phosphorescent Bay, Puerto Rico	1961	McLaughlin and Zahl (1961) includes soil extract	24-28°C	3 x 40 W fluorescent	14:10-hour L:D cycle	30 days	> 6 months	Mc Laughlin and Zahl (1961)
var. *bahamense*	Oyster Bay, Jamaica	1960-1979	Sweeney and Hastings (1957) includes soil extract	15-37°C 25°C optimum	1 x 20 W tungsten				
var. *bahamense*	Puerto Rico		F/10, K/10 with and without local soil extract; Erdschreiber	24-25°C	5,400 lux high yield cool white fluorescent	12:12-hour L:D cycle	transfer to density > 700 cells/ ml	19 years	Biggley et al. (1969), Seliger, H. (pers. comm.)
var. *compressum*	Palau	May 1982 May 1983	modified von Stosch's medium and soil extract	30 ± 2°C	150 $\mu Em^{-2} s^{-1}$ cool white fluorescent	16:8-hour L:D cycle	-	several years (lost by incu-bator failure)	Steidinger, K. (pers. comm.)
					3,000-5,000 lux	18:6-hour L:D cycle	28 days	4 years (lost by incubator failure)	Oshima et al. (1985)

Table 2. Composition of culture media tested for *Pyrodinium bahamense* var. *compressum* (after Oshima et al. 1985).

		A	B	C	D	E	F	G	H	I
Seawater*	(ml)	1,000	1,000	1,000	750	1,000	1,000	1,000	900	1,000
Distilled water	(ml)	-	-	-	250	-	-	-	0-90	-
$NaNO_3$	(mg)	100	43	-	-	150	70	-	-	14
KNO_3	(mg)	-	-	72	202	-	-	100	-	-
$Na_2HPO_4 \cdot 12H_2O$	(mg)	20	-	-	-	-	-	-	-	-
$NaH_2PO_4 \cdot H_2O$	(mg)	-	5	-	-	10	-	-	-	1.6
K_2HPO_4	(mg)	-	-	35	-	-	-	10	-	-
KH_2PO_4	(mg)	-	-	4.5	-	-	-	-	-	-
Na_2-glycerophosphate	(mg)	-	-	10.5	-	-	10	-	-	-
$Na_2SiO_3 \cdot 9H_2O$	(mg)	-	-	-	-	30	-	-	-	-
$FeCl_3$	(µg)	-	540	-	970	-	240	100	-	300
Fe (as EDTA)	(µg)	-	56	500	-	-	500	-	-	130
Ferric sequestrene	(mg)	-	-	-	-	10	-	-	-	40
Na_2 EDTA	(mg)	-	3.7	-	10	-	5.9	-	-	8
$MnCl_2 \cdot 4H_2O$	(µg)	-	240	-	75	360	-	-	-	-
$CuSO_4 \cdot 5H_2O$	(µg)	-	-	-	-	19.6	-	-	-	-
$ZnSO_4 \cdot 7H_2O$	(µg)	-	-	-	-	44	55	-	-	-
$CoCl_2 \cdot 6H_2O$	(µg)	-	-	-	-	20	10	-	-	-
Na_2MoO_2	(µg)	-	-	-	-	12.6	-	-	-	-
H_3BO_3	(µg)	-	-	-	-	-	500	-	-	-
Vitamin B_{12}	(µg)	-	1	10	10	1	2	-	1	-
Biotin	(µg)	-	-	5	-	1	1	-	1	-
Thiamine HC1	(µg)	-	-	500	-	200	100	-	10	-
Soil extract	(ml)	50	-	-	-	-	-	20	10-100	50
Tris (hydroxymethyl)-aminomethane	(mg)	-	-	500	-	-	100	-	-	100
Yeastlate	(mg)	-	-	-	-	-	-	-	1	-
Antibiotic sol.	(ml)	-	-	-	-	-	-	-	6.5	1
pH		-	-	8.2	-	-			7.8	8.0

Common names of media and references:

A: Fϕyn (1934) Erdschreiber
B: von Stosch (1964)
C: Iwasaki (1981) medium SW-II
D: Sweeney (1954)
E: Guillard and Ryther (1962) medium F
F: Provasoli (1968) medium ES-I
G: Barker (1935)
H: McLaughlin and Zahl (1961)
I: modified von Stosch's medium (Oshima et al. 1985)

* All the media were prepared with filtered seawater taken at Shizugawa Bay, Miyagi Prefecture.

media have been most successful. McLaughlin and Zahl (1961) employed acid hydrolyzed mud from Phosphorescent Bay with added vitamins and yeast autolysate for var. *bahamense* while the soil-based medium of Sweeney and Hastings (1957) was used by Biggley et al. (1969).

Oshima et al. (1985) carried out detailed comparative growth studies with eight different media (Table 2). *P. bahamense* from Palau (var. *compressum*) died within five days in the medium used by McLaughlin and Zahl (1961) for var. *bahamense*. It also failed to survive in the media F (Guillard and Ryther 1962) and ES-I (Provasoli 1968) and the media used by Barker (1935) and Sweeney (1954). In von Stosch's (1984), Erdschreiber (Fϕyn 1934) and Iwasaki's (1961) SW-II media, the organism survived for five days and the survival time could be prolonged by lowering the concentration of the nutrients (Table 3). Division of cells was observed in 1/3 strength von Stosch's medium but the medium could not maintain growth and all the cells died after 25 days. Lowering the

Table 3. Comparative growth of *Pyrodinium bahamense* var. *compressum* in different culture media (after Oshima et al. 1985).

Culture medium	Strength[a]	Seawater conc. (%)	Result[b]
Fϕyn's Erdschreiber (A)	1	100	+
	1/5	100	++
	1/10	100	++++
	1/10	90	++
	1/10	80	+
von Stosch (B)	1	100	+
	1/5	100	+++
	1/10	100	++++
Iwasaki's SW-II (C)	1	100	++
	1	80	++
Sweeney (D)	1	100	-
	1	75	-
Guillard's F (E)	1	100	-
Provasoli's ES-I (F)	1	100	-
Barker (G)	1	100	-
McLaughlin (H)	1	100	-

[a]Concentration of nutrients shown in Table 3 is expressed as in Table 2.

[b]Survival time: less than 5 days (-), more than 5 days (+), 10 days (++), 15 days (+++), and 20 days (++++).

salinity by diluting with distilled water was also not effective for growth. Subsequently, extracts of soils taken from four different localities in Sendai, Japan, were added to the media in concentrations of 5-500 mg wet weight/l. Vigorous growth was observed in modified von Stosch's medium by addition of a soil extract (50 mg/l) (Table 2).

Steidinger (pers. comm.) also had success in culturing var. *bahamense* from Puerto Rico in Erdschreiber medium as well as F/10 (Guillard 1975) and K/10 (Keller and Guillard 1985) media, using Gulf of Mexico water as a base. Dilute soil extract was added in some cases.

The spatial association between *Pyrodinium* and mangrove areas (Seliger, this vol.) suggests that mangrove communities might provide necessary growth factors similar to those in soil extract.

Antibiotics. In order to obtain axenic cultures, Oshima et al. (1985) included streptomycin (1 mg/l seawater) in the modified von Stosch's medium. However, lack of bacterial flora may have contributed to difficulties in culturing and bacteria isolated with cells may in fact facilitate the culture of *P. bahamense.* Antibiotics were also employed by McLaughlin and Zahl (1961) to produce two axenic culture lines.

Temperature. A wide temperature tolerance of 15-37°C is shown by *P. bahamense* in culture (Seliger, pers. comm.; Table 1) and most cultures have been kept at 24-30°C.

Salinity. Seawater medium at full-strength salinity was found to be more effective for growth than dilutions of 75 to 90% (Oshima et al. 1985). All other culture attempts have used full-strength seawater.

pH. A significant effect of pH of modified von Stosch's medium was observed for *P. bahamense* var. *compressum* (Fig. 1; Oshima et al. 1985). In a 20-day-old culture inoculated at 40 cells/ml cell density varied between 0.9 and $1.8.10^3$ cells/ml over a pH range 7.0-8.5 with the maximum density at a pH of 8.

Illumination. Light intensity of 3,000-5,000 lux was employed by Oshima et al. (1985). To simulate tropical conditions a 12:12-hour light:dark cycle would be appropriate, but a 14:10-hour light:dark cycle (McLaughlin and Zahl 1961) and a 18:6-hour light:dark cycle (Oshima et al. 1985) have also been used successfully. The onset of the light period was significant in promoting cell division of var. *bahamense in situ* in Oyster Bay (Buchanan 1968). This may also apply to cultures.

Transfer of cultures. The size of inoculum was found to be critical in the maintenance of cultures (Seliger, pers. comm.). An initial density of at least 700 cells/ml was necessary to establish a vigorous culture. Similarly, when a seed culture with a density more than

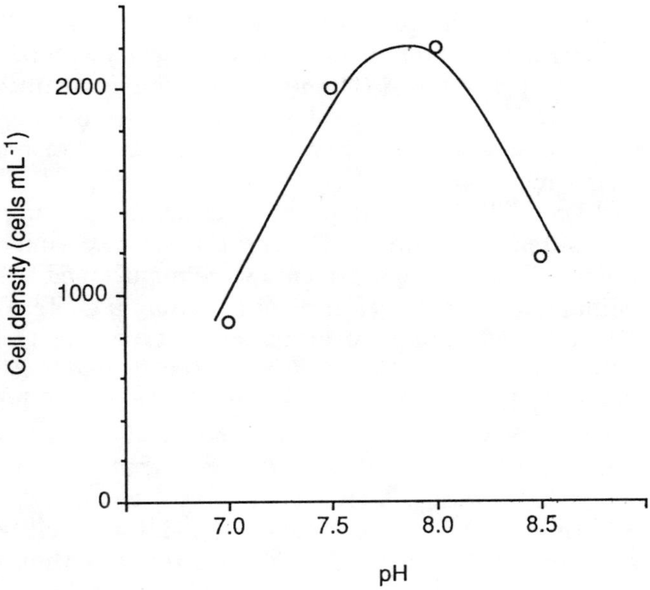

Fig. 1. Effect of pH on the growth of *Pyrodinium bahamense* var. *compressum*. Cell density at 20 days (inoculum 40 cells/ml) as a function of pH of modified von Stosch's medium (after Oshima et al. 1985).

1,000 cells/ml was used Oshima et al. (1985) could grow var. *compressum* under suboptimal conditions in F medium without soil extract. This may be due to growth factors in the medium transferred from the parent culture. A 30-day transfer period was employed both by McLaughlin and Zahl (1961) and Oshima et al. (1985) with no loss of vigor in the cultures.

Cyst Formation and Germination

The resting of cyst of var. *bahamense* was identified by Wall and Dale (1969) from germination experiments of cysts found in Bermuda and Puerto Rico surface sediments. As for other dinoflagellates, the resting cyst of *P. bahamense* is probably produced as part of a sexual life cycle (Pfiester and Anderson 1987) but the conditions of resting cyst formation for this species are still unknown and no resting cysts have yet been observed in culture.

The temperature at which sediment samples or resting cysts are maintained may be critical to resting cyst viability and germination. Whereas the germination of cysts from temperate waters is often enhanced by a period of "conditioning" at low temperature, e.g. 5°C (Anderson and Wall 1978), tropical dinoflagellates are unlikely to experience low temperatures which thus may be detrimental to survival and germination.

Wall and Dale (1969) kept Bermuda sediments at a temperature of 15.5°C for one to two months before germination experiments were carried out, while Puerto Rico sediments were maintained at 26°C, in both cases similar to the temperatures at the sampling site. Wall and Dale (1969) conducted successful germination experiments at 26°C for the Bermuda cysts and at 27°C for those from Puerto Rico, using a 14:10-hour light:dark cycle and illumination of 5,000-8,000 lux. These authors found that resting cysts collected in February did not germinate in mid-March under conditions that did result in germination several weeks later. These results suggest that *P. bahamense* resting cysts have a dormancy period. Germination of var. *bahamense* resting cysts liberated either motile cells that secreted thecae, or nonmotile, spherical to ovoid cells. Cultures were not established from these germination studies. Asexual cyst stages or temporary cysts and athecate vegetative stages were also produced in incubation studies by Buchanan (1969) using natural samples from a var. *bahamense* bloom in Oyster Bay. Round temporary cysts with a hyaline wall formed both within the theca and by rounding up of a gymnodinioid stage.

Conclusions and Recommendations

While *Pyrodinium bahamense* has been successfully isolated and maintained in culture several times, its fragile nature and incubator failures have led to the eventual loss of all these cultures. For culturing of *P. bahamense* it is recommended that:

1) Cell isolations should be made from blooms or red tides.
2) Seawater from the collection locality should be used for making media.
3) Cells may need to be isolated into small culture vessels or micropipettes in petri dishes.
4) Soil extract may be an important growth promoter in the medium.
5) Antibiotics may be deleterious to growth.
6) A culture medium of pH 8 is optimal for growth.

7) Temperature in the range 24-30°C is preferable with limits of 15-37°C.
8) A tropical photoperiod may be important to promote growth.

The varieties *compressum* and *bahamense* form two geographically separated populations and differ in cell morphology and chain formation. Strains of var. *bahamense* from Latin America and var. *compressum* from the Indo-West Pacific need to be cultured from as many localities as possible and characterized taxonomically, biochemically and physiologically.

References

Anderson, D.M. and D. Wall. 1978. Potential importance of benthic cysts of *Gonyaulax tamarensis* and *G. excavata* in initiating toxic dinoflagellate blooms. J. Phycol. 14: 224-234.

Barker, H.A. 1935. The culture and physiology of marine dinoflagellates. Arch. Mikrobiol. 6: 171-181.

Biggley, W.H., E. Swift, R.J. Buchanan and H.H. Seliger. 1969. Stimulable and spontaneous bioluminescence in the marine dinoflagellates, *Pyrodinium bahamense, Gonyaulax polyedra* and *Pyrocystis lunula*. J. Gen. Physiol. 54: 96-122.

Buchanan, R.J. 1968. Studies at Oyster Bay in Jamaica, West Indies. IV. Observations on the morphology and asexual cycle of *Pyrodinium bahamense* Plate. J. Phycol. 4: 272-277.

Fφyn, B. 1934. Lebenszyklus, Cytologie and Sexualität der Chlorophycee *Cladophora suhriana* Kützing. Arch. Protistenk. 83: 1-56.

Guillard, R.L. 1975. Phytoplankton for feeding marine invertebrates, p. 29-60. *In* W.L. Smith and M.H. Chanley (eds.) Culture of marine invertebrate animals. Plenum Press, New York.

Guillard, R.L. and J.H. Ryther. 1962. Studies of marine planktonic diatoms 1. *Cyclotella nana* Hustedt and *Detonula confervacea* (Cleve) Gran. Can. J. Microbiol. 8: 229-239.

Iwasaki, H. 1961. The life-cycle of *Porphyra tenera* in vitro. Biol. Bull. 121: 173-187.

Keller, M.D. and R.R.L. Guillard. 1985. Factors significant to marine dinoflagellate culture, p. 113-116. *In* D.M. Anderson, A.W. White and D.G. Baden (eds.) Toxic dinoflagellates, Elsevier, New York.

McLaughlin, J.J.A. and P.A. Zahl. 1961. In vitro culture of *Pyrodinium*. Science 134: 1878.

Oshima, Y., T.S. Harada and T. Yasumoto. 1985. Paralytic shellfish poisoning, p. 20-31. *In* T. Yasumoto (ed.) Studies on tropical fish and shellfish infested by toxic dinoflagellates. Faculty of Agriculture, Tohuku University.

Pfiester, L.A. and D.M. Anderson. 1987. Dinoflagellate reproduction, p. 611-648. *In* F.J.R. Taylor (ed.) The biology of dinoflagellates. Botanical Monographs, Vol. 21. Blackwell Scientific Publications, Oxford.

Provasoli, L. 1968. Media and prospects for the cultivation of marine algae, p. 63-75. *In* A. Watanabe and A. Hattori (eds.) Cultures and collections of algae. Proc. U.S.-Japan Conf. Hakane, Sept. 1966. Jap. Soc. Plant Physiol.

Sweeney, B.M. 1954. *Gymnodinium splendens,* a marine dinoflagellate requiring vitamin B_{12}. Am. J. Bot. 41: 821-824.

Sweeney, B.M. and Hastings. 1957. Characteristics of the diurnal rhythm of luminescence in *Gonyaulax polyedra*. J. Cell Comp. Physiol. 49: 115-128.

von Stosch, H.A. 1964. Wirkungen von Jod und Arsenit auf Meeresalgen in Kultur, p. 142-150.. *In* A.D. de Virville and J. Feldmann (eds.) Proc. IV. Int. Seaweed Symp. Pergamon Press.

Wall, D. and B. Dale. 1969. The "hystrichosphaerid" resting spore of the dinoflagellate *Pyrodinium bahamense*, Plate, 1906. J. Phycol. 5: 140-149.

Appendix

1. For general algal culture techniques the following references are useful.

    Guillard, R.R.L. 1975. Culture of phytoplankton for feeding marine invertebrates, p. 29-60. *In* W.L. Smith and M.H. Chanley (eds.) Culture of marine invertebrates animals, Plenum, New York.

    Stein, J.R. 1973. Handbook of phycological methods: culture methods and growth measurements. Cambridge University Press, London and New York.

    Ukeles, R. 1976. Cultivation of unicellular plants. *In* O. Kinne (ed.) Marine ecology vol. III: cultivation methods. John Wiley, London.

2. For culturing dinoflagellates the following references should be consulted.

    Guillard, R.R.L. and L.E. Brand. 1985. Culturing, p. 538-539. *In* D.M. Anderson, A.W. White and D.G. Baden (eds.) Toxic dinoflagellates, Elsevier, New York.

    Guillard, R.R.L. and M.D. Keller. 1984. Culturing dinoflagellates, p. 391-442. *In* D.L. Spector (ed.) Dinoflagellates, Academic Press, London.

    Keller, M.D. and R.R.L. Guillard. 1985. Factors significant to marine dinoflagellate culture, p. 113-116. *In* D.M. Anderson, A. White and D.G. Baden (eds.) Toxic dinoflagellates, Elsevier, New York.

    Siegelman, H. and M. Levandowski. 1979. Culturing of dinoflagellates, p. 471-472. *In* D.L. Taylor and H.H. Seliger (eds.) Toxic dinoflagellate blooms, developments in marine biology vol. I, Elsevier, North Holland.

Standard Mouse Bioassay for Paralytic Shellfish Toxins

The following standard method has been copied with permission of the Official Methods of Analysis, 1990. William Horwitz, Ed., Association of Official Analytical Chemists, Washington, DC, USA, pages 881-882.

Paralytic Shellfish Poison Biological Method

(*Caution:* Use rubber gloves when handling materials which may contain paralytic shellfish poison).

Materials

(a) Paralytic shellfish poison (saxitoxin) standard solution - 100 µg/ml. Available from Division of Contaminants Chemistry, Natural Products and Instrumentation Branch (HFF-423), Food and Drug Administration, 200 C St., SW, Washington, DC 20204, as acidified 20% alcohol solution. Standard is stable indefinitely in cool place.

(b) Paralytic shellfish poison working standard solution - 1 µg/ml. Dilute 1 ml standard solution to 100 ml with distilled water. Solution is stable several weeks at 3-4°C.

(c) Mice - Healthy mice, 19-21 g, from stock colony used for routine assay. If <19 g or >21 g, apply correction factor to obtain true death time (see Sommer's Table). Do not use mice weighing 23 g and do not re-use mice.

Standardization of Bioassay

Dilute 10 ml aliquots of 1 µg/ml standard solution with 10, 15, 20, 25 and 20 ml water, respectively, until intraperitoneal injection of 1 ml doses into few test mice causes median death time of 5-7 min.

pH of dilutions should be 2-4 and must not be >4.5. Test additional dilutions in 1 ml increments of water, e.g., if 10 ml diluted with 25 ml water kills mice in 5-7 min, test solutions diluted 10 + 24 and 10 + 26.

Inject group of 10 mice with each of 2 or preferably 3 dilutions that fall within median death time of 5-7 min. Give 1 ml dose to each mouse by intraperitoneal injection and determine death time as time elapsed from completion of injection to last gasping breath of mouse.

Repeat assay 1 or 2 days later, using dilutions prepared above which differed by 1 ml increments of water. Then repeat entire test, starting with testing of dilutions prepared from newly prepared working standard solution.

Calculate median death time for each group of 10 mice used on each dilution. If all groups of 10 mice injected with any 1 dilution gave median death time <5 or >7 min, disregard results from this dilution in subsequent calculations. On the other hand, if any groups of 10 mice injected with 1 dilution gave median death time falling between 5 and 7 min, include all groups of 10 mice used on that dilution, even though some of median death times may be <5 or >7 min. From median death time for each group of 10 mice in each of selected dilutions, determine number of mouse units/ml from Sommer's Table. Divide calculated μg poison/1 ml by mouse units/1 ml to obtain conversion factor (CF value) expressing μg poison equivalent to 1 mouse unit. Calculate average of individual CF values, and use this average value as reference point to check routine assays. Individual CF values may vary significantly within laboratory if techniques and mice are not rigidly controlled. This situation will require continued use of working standard or secondary standard, depending on volume of assay work performed.

Use of Standard with Routine Assays of Shellfish

Check CF value periodically as follows: If shellfish products are assayed less than once a week, determine CF value on each day assays are performed by injecting 5 mice with appropriate dilution of working standard. If assays are made on several days during week, only 1 check need to be made each week on dilution of standard such that median death time falls within 5-7 min. CF value thus determined should check with average CF value within ± 20%. If it does not check within this range, complete group of 10 mice by adding 5 mice to the 5 mice already injected, and inject second group of 10 mice with same dilution of standard. Average CF value determined for second group with that of first group. Take resulting value as new

CF value. Variation of >20% represents significant change in response of mice to poison, or in technique of assay. Changes of this type require change in CF value.

Repeated checks of CF value ordinarily produce consistent results within ± 20%. If wider variations are found frequently, the possibility of uncontrolled or unrecognized variables in method should be investigated before proceeding with routine assays.

Preparation of Sample

(a) Clams, oysters and mussels. - Thoroughly clean outside of shellfish with fresh water. Open by cutting adductor muscles. Rinse inside with fresh water to remove sand or other foreign material. Remove meat from shell by separating adductor muscles and tissue connecting at hinge. Do not use heat or anesthetics before opening shell, and do not cut or damage body of mollusk at this state. Collect about 100-150 g meats in glazed dish. As soon as possible, transfer meats to No. 10 sieve without layering, and let drain 5 min. Pick out pieces of shell and discard drainings. Grind in household-type grinder with 1/8-1/4" (3-6 mm) holes or in blender until homogeneous.

(b) Scallops. - Separate edible portion (adductor muscle) and apply test to this portion alone. Drain and grind as in (a).

(c) Canned shellfish. - Place entire contents of can (meat and liquid) in blender and blend until homogeneous or grind 3 times through meat chopper. For large cans, drain meat 2 min on No. 8-12 sieve and collect all liquid. Determine weight of meat and volume of liquid. Recombine portion of each in proportionate amounts. Blend recombined portions in blender (or grind) until homogeneous.

Extraction

Weigh 100 g well mixed material into tared beaker. Add 100 ml 0.1 *N* HCl, stir thoroughly and check pH. (pH should b < 4.0, preferably about 3.0. If necessary, adjust pH as indicated below.) Heat mixture, boil gently 5 min, and let cool to room temperature. Adjust cooled mixture to pH 2.0-4.0 (never > 4.5) as detected by *BHD Universal Indicator, phenol blue, Congo red paper,* or pH meter. To lower pH, add 5 *N* HCl dropwise with stirring; to raise pH, add 0.1 *N* NaOH dropwise with constant stirring to prevent local alkalinization and consequent destruction of poison. Transfer mixture to graduated cylinder and dilute to 200 ml.

Return mixture to beaker, stir to homogeneity, and let settle until portion of supernate is translucent and can be decanted free of solid particles large enough to block 26-gauge hypodermic needle. If necessary, centrifuge mixture or supernate 5 min at 3,000 rpm or filter through paper. Only enough liquid to perform bioassay is necessary.

Mouse Test

Intraperitoneally inoculate each test mouse with 1 ml acid extract. Note time of inoculation and observe mice carefully for time of death as indicated by last gasping breath. Record death time from stopwatch or clock with sweep second hand. One mouse may be used for initial determination, but 2 or 3 are preferred. If death time or median death time of several mice is < 5 min, make dilution to obtain death times of 5-7 min. If death time of 1 or 2 mice injected with undiluted sample is >7 min, a total of ≥3 mice must be inoculated to establish toxicity of sample. If large dilutions are necessary, adjust pH of dilution by dropwise addition of dilute HC1 (0.1 or 0.01 *N*) to pH 2.0-4.0 (never > 4.5). Inoculate 3 mice with dilution that gives death times of 5-7 min.

Calculation of Toxicity

Determine median death times of mice, including survivors, and from Sommer's Table determine corresponding number of mouse units. If test animals weigh < 19 g or > 21 g, make correction for each mouse by multiplying mouse units corresponding to death time for that mouse by weight correction factor for that mouse from Sommer's Table; then determine median mouse unit for group. (Consider death time of survivors as > 60 min or equivalent to < 0.875 mouse unit in calculating median). Convert mouse units to μg poison/ml by multiplying by CF value.

μg Poison/100 g meat

= (μg/ml) x dilution factor x 200

Consider any value > 80 μg/100 g as hazardous and unsafe for human consumption.

Death time: mouse unit relations for paralytic shellfish poison (acid).

Death time[a]	Mouse units	Death time[a]	Mouse units
1:00	100	5:00	1.92
10	66.2	05	1.89
15	38.3	10	1.86
20	26.4	15	1.83
25	20.7	20	1.80
30	16.5	30	1.74
35	13.9	40	1.69
40	11.9	45	1.67
45	10.4	50	1.64
50	9.33		
55	8.42	6:00	1.60
		15	1.54
2:00	7.67	30	1.48
05	7.04	45	1.43
10	6.52		
15	6.06	7:00	1.39
20	5.66	15	1.35
25	5.32	30	1.31
30	5.00	45	1.28
35	4.73		
40	4.48	8:00	1.25
45	4.26	15	1.22
50	4.06	30	1.20
55	3.88	45	1.18
3:00	3.70	9:00	1.16
05	3.57	30	1.13
10	3.43		
15	3.31	10:00	1.11
20	3.19	30	1.09
25	3.08		
30	2.98	11:00	1.075
35	2.88	30	1.06
40	2.79		
45	2.71	12:00	1.05
50	2.63		
55	2.56	13	1.03
		14	1.015
4:00	2.50	15	1.000
05	2.44	16	0.99
10	2.38	17	0.98
15	2.32	18	0.972
20	2.26	19	0.965
25	2.21	20	0.96
30	2.16	21	0.954
35	2.12	22	0.948
40	2.08	23	0.942
45	2.04	24	0.937
50	2.00	25	0.934
55	1.96	30	0.917
		40	0.898
		60	0.875

[a]Minutes: seconds.

Correction table for weight of mice.

Wt of mice, g	Mouse units
10	0.50
10.5	0.53
11	0.56
11.5	0.59
12	0.62
12.5	0.65
13	0.675
13.5	0.70
14	0.73
14.5	0.76
15	0.785
15.5	0.81
16	0.84
16.5	0.86
17	0.88
17.5	0.905
18	0.93
18.5	0.95
19	0.97
19.5	0.985
20	1.000
20.5	1.015
21	1.03
21.5	1.04
22	1.05
22.5	1.06
23	1.07

Analysis of *Pyrodinium bahamense* PSP Toxins by High-Performance Liquid Chromatography

YASUKATSU OSHIMA
Faculty of Agriculture
Tohoku University
1-1 Tsutsumidori Amamiyamachi
Sendai 981, Japan

OSHIMA, Y. 1989. Analysis of *Pyrodinium bahamense* PSP toxins by high-performance liquid chromatography, p. 273-278. *In* G.M. Hallegraeff and J.L. Maclean (eds.) Biology, epidemiology and management of *Pyrodinium* red tides. ICLARM Conference Proceedings 21, 286 p. Fisheries Department, Ministry of Development, Brunei Darussalam, and International Center for Living Aquatic Resources Management, Manila, Philippines.

Analysis of PSP toxins by mouse bioassay has several disadvantages such as (1) insensitivity (detection limit 1.0 MU (mouse unit)/ml); (2) imprecision (20% error); (3) the need to maintain a mouse colony; and (4) ethical objections against animal experiments (in some countries). Therefore, alternative chemical procedures have been developed in which PSP toxin fractions are being separated by high-performance liquid chromatography (HPLC), oxidized to fluorescent derivatives and then detected by fluorescence. For the accurate separation of complex toxin mixtures (e.g., from *Gymnodinium catenatum*), three different isocratic chromatographic separations are required to distinguish all 18 known toxin fractions (Oshima et al. 1988). Fortunately, toxin profiles derived from *Pyrodinium bahamense* appear to be relatively simple and contain only STX, neoSTX, dcSTX, GTX5 and sometimes GTX6 (Oshima, this vol.). This has formed the basis for the following simplified HPLC method which uses a 20-minute isocratic elution with a single solvent system.

Preparation of Extracts

Prepare a shellfish meat or fish stomach extract according to the AOAC method for standard mouse bioassay (0.1 N HCl, heating for 5 min.). Centrifuge the extract at 10,000 g for 5-10 min. Wash a Sep-Pak C-18 cartridge column (Waters) with 10 ml of methanol and drain; equilibrate the column with 10 ml of distilled water and drain again. Pass the extract through the cartridge column, discard the first 1.5 ml of eluate and collect 0.5 ml. into a tube fitted with an ultrafiltration filter (Millipore Ultrafree C3GC, 10,000 dalton cut-off). Centrifuge the sample at 5,000 g for 5 min.

Apparatus (Fig. 1)

(a) High-performance liquid chromatograph. Any model capable to flow a mobile phase at more than 1.0 ml/min or 200 kg/cm^2 with reliable flow control.

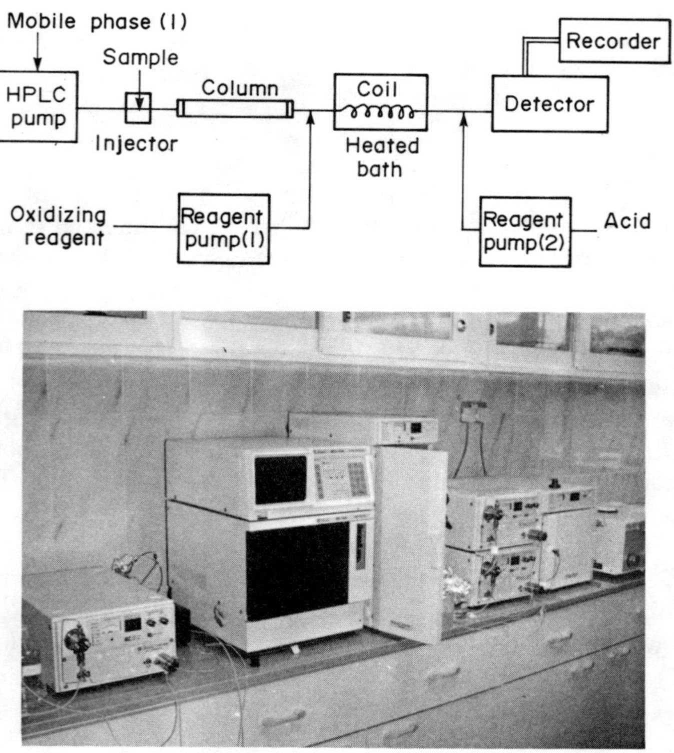

Fig. 1. Flow diagram for HPLC system (upper) and the apparatus used during the Brunei Darussalam workshop (lower).

(b) A loop or variable volume (syringe loading) injector (10-100 μl).

(c) Chromatographic column. Well capped chemically bonded C8 silicagel of column size 4.6 mm i.d. x 150 mm long or 4.6 x 250 mm (e.g., Develosil C8-5, Nomura Chemical or Inertsil C8 Gasukuro Kogyo)

(d) Post column reaction system: (1) pump for delivering the oxidizing reagent, preferably a smaller pulse model with acid resistant pump head; (2) reaction coil of Teflon tubing, 0.5 mm i.d. and 10 m long; (3) water bath to keep reaction coil at 65 ± 1°C; (4) pump for delivering acid to acidify the reaction mixture (as in 1).

(e) Highly sensitive fluorescence detector with a low volume (12 μl) HPLC flow cell, a 150-W Xenon lamp as light source and capable of selecting excitation and emission wavelengths of 330 and 390 nm, respectively, by dual monochromators.

(f) Recorder. Toxin concentrations can be determined by measuring peak heights from chromatograms taken by conventional 1 pen- recorder fitted to the detector output (usually 10 mV). Electronic integrators are more convenient in covering a wider range of output and for more accurate calculation of peak area.

Stock Solutions

(1) 100 nM sodium 1-heptanesulfonate solution. Dissolve 2.02 g of HPLC-grade reagent in 100 ml of distilled water.

(2) 500 mM phosphoric acid. Dissolve 28.8 g of concentrated analytical grade (85%) phosphoric acid in distilled water to make up 500 ml.

(3) 1 N ammonium hydroxide. Dilute concentrated (25%) ammonia water with 12 times distilled water.

(4) 350 nM periodic acid. Dissolve 7.98 g of analytical grade reagent in 100 ml of distilled water.

(5) 250 mM disodium phosphate. Dissolve 44.77 g of analytical grade $Na_2HPO_4.2H_2O$ in distilled water and make up to 500 ml.

(6) 1 N sodium hydroxide. Dissolve 4 g of NaOH in 100 ml of distilled water.

Mobile Phase

Dissolve 10 ml of stock solution 1 and 30 ml of solution 2 in 450 ml of distilled water, adjust to pH 7.1 by adding NH_4OH solution 3 and make up to 500 ml. Add 30 ml of acetonitrile, mix well and de-gas by sonication.

Oxidizing Reagent

Prepare fresh reagent daily by adding 10 ml of stock solution 4 and 100 ml of solution 5 in 100 ml of distilled water. Titrate to pH 9.0 with sodium hydroxide solution 6 and dilute to 500 ml with distilled water.

Operation of the HPLC

Start-up procedure

(1) Prime the HPLC pump with the mobile phase and establish a flow rate of 0.8 ml/min. Run the HPLC pump for at least 15 min.
(2) Prime the post column reaction pumps with oxidizing reagent and 0.5 M acetic acid, respectively. Establish reagent flow rates of 0.4 ml/min. and a reaction coil temperature of 65°C.
(3) Operate the detector at excitation wavelength 330 nm and emission wavelength 390 nm.
(4) Inject a standard toxin solution repeatedly until retention times are constant, indicating that the column has been equilibrated with mobile phase.

Sample analysis

(1) Inject 5-10 µl of extract and record peak height or peak area
(2) Inject a standard toxin solution after every 4 samples to ensure that all systems are working properly
(3) Use the nearest standard chromatogram for calculation of toxin concentrations and use the potency values in Table 1 for calculation of the toxicity of the sample.

Shut-down procedure

(1) Switch off recorder, detector and water bath.
(2) Prime reaction pumps with distilled water for at least 15 min. and stop.
(3) After running the reaction pumps for 5 min. prime the HPLC pump with 25% acetonitrile and flow for at least 10 min. Then

change the solvent to 50% acetonitrile and run for another 30 min.

(4) Change mobile phase to 25% acetonitrile and run the pump for 10 min. and stop.

Trouble Shooting

The degree of fluorescence response for the individual toxin fractions varies according to oxidation condition which is dependent on periodate concentration, reaction temperature, pH and time. It is important to determine optimum conditions for your own system using known toxin standards. Typical detection limits that can be achieved are summarized in Table 1 and typical chromatograms are shown in Fig. 2. A number of factors affect the retention time of the different toxin fractions. Decreasing the ion-pairing reagent concentration, increasing the buffer concentration and increasing the acetonitrile concentration all tend to speed up elution of the toxins. Change of pH of the mobile phase also affects retention times.

Table 1.

Toxin fraction	Detection limit (10^{-3} MU/ml)	Mouse intra-peritoneal potency (MU/μmole)
STX	19	2,045
neoSTX	105	1,038
dcSTX	35	1,220
GTX 5	1.5	350
(GTX 6)	10	180

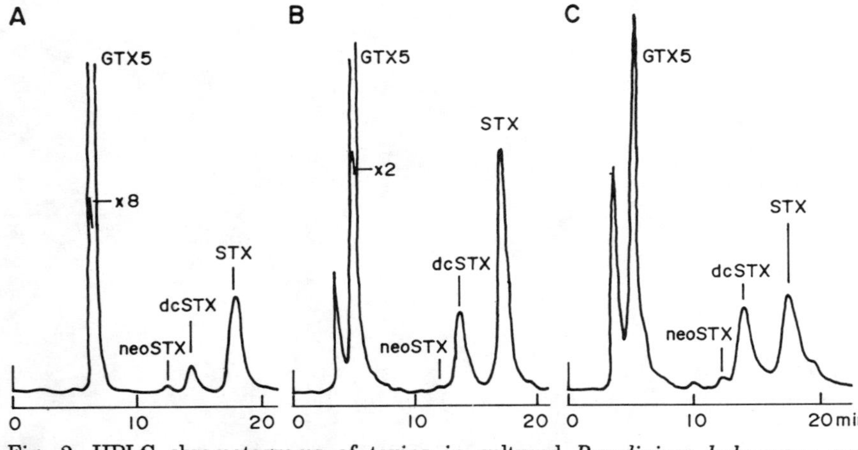

Fig. 2. HPLC chromatograms of toxins in cultured *Pyrodinium bahamense* var. *compressum* from Palau (A), Philippine mussels (B) and Solomon Islands rock oyster (C) (after Oshima et al. 1987).

References

Oshima, Y., T. Yasumoto, G. Hallegraeff and S. Blackburn. 1987. Paralytic shellfish toxins and causative organisms in the tropical Pacific and Tasmanian waters, p. 423-428. *In* P. Gopalakrishnakone and C.T. Tan (eds.) Progress in venom and toxin research. National University of Singapore, Singapore.

Oshima, Y., K. Sugino and T. Yasumoto. 1988. Latest advances in HPLC analysis of paralytic shellfish toxins, p. 319-326. *In* S. Natori, K. Hashimoto and Y. Ueno (eds.) Mycotoxins and phycotoxins '88. Elsevier, New York.

Guidelines in Investigating PSP Epidemics

NIÑO ISMAEL S. PASTOR
MARK E. WHITE
MANUEL M. DAYRIT
Field Epidemiology Training Program
Department of Health
Philippines

Introduction

General objectives. The purpose of investigating Paralytic Shellfish Poisoning (PSP) epidemics is to minimize illnesses and prevent further deaths.

Special objectives. To achieve the prime directive, the following specific objectives have to be attained (CDC 1987):

1. Detection of PSP
2. Verifying the diagnosis of PSP
3. Confirming the existence of an epidemic
4. Characterizing the epidemic
5. Identifying the cause of the epidemic
6. Implementing control measures

Detection of Epidemics

Who gives you information about PSP occurrence? Before epidemics can be investigated, PSP occurrences have to be known. Three sources of information reveal the occurrence of PSP, namely (a) surveillance system, (b) medical reports, and (c) other reports (Bres 1986).

Data from surveillance systems are available from most departments or ministries of health as bulletins, or newsletters issued at regular intervals. PSP is reported under foodborne disease.

Most countries require physicians and health workers to report foodborne diseases to the Ministry or Department of Health.

Other sources of PSP incidents include: media, local government officials, and concerned citizens.

Determine the validity of the report. Not all accounts of PSP are true, therefore the validity of a report must be determined. When two out of the three sources of information are consistent, something must be happening in the field requiring further investigation.

Verifying the Diagnosis

When a report is valid, the next step is to verify the type of illness the patients have. The diagnosis at this stage does not have to be final. Remember, you are out on the field with no fancy equipment at hand. Any previously healthy person who suddenly develops at least two motor and two sensory symptoms (Table 1) is suspected to be afflicted with PSP.

Inquiries regarding the patients symptoms can be made by telephone or on-site inspection. The decision to do on-site verification depends upon: (a) the distance of the epidemic site and (b) the availability of resources.

Table 1. Symptoms of PSP.

Symptoms		
Gastrointestinal	Motor abnormalities	Sensory abnormalities
Vomiting	Inability to ambulate	Numbness
Abdominal pain	Paresis of extremities	General body malaise
Water diarrhea	Dyspnea	Dizziness
Nausea	Dysphagia	Light headed sensation
Hypersalivation	Diplopia	Headache
	Paralysis	Paresthesia
	Dysphonia	Felt hot
		Dysthesia
		Short tongue sensation
		Pruritus

Is There An Epidemic?

When PSP is suspected, the next step is to determine whether there is an epidemic.

PSP epidemics are suspected when 2 or more previously healthy persons (MacDonald and Griffin 1982) complain of at least two motor and two sensory symptoms (Pastor et al., this vol.) after sharing a common meal (Table 1).

Describe the Patients by Time, Place and Person

The first step in understanding an epidemic is to describe it. The simplest and most useful way to do this is to list the patients by time, place and person (see Appendix I).

To describe an epidemic by time, collect the dates at which people first became sick, the onset date. Then plot these on a graph, in an epidemic curve or epidemic histogram. The number of patients is plotted on the Y-axis, and days, weeks or months are plotted on the X-axis. Useful clues can be derived about how PSP was transmitted among the patients. It also measures the efficacy of the control measures imposed. The latter is illustrated in the example given below.

A newspaper reported an epidemic of PSP 150 km southwest of the Department of Health. Since it was near, investigators visited the site the following day. In the hospital, the investigators validated the newspaper report and confirmed the diagnosis of PSP. Further inquiries indicated that there were more than two previously healthy persons with more than two motor and sensory symptoms who shared a meal of green mussels. It was concluded that indeed an epidemic existed. After a week, the investigators imposed a ban on shellfish gathered from the site where the patients, who were mostly fishermen, gathered the mussels. A month later cases continued to occur. The investigators concluded that the ban was not effective. A different strategy was adopted. This illustrates how the epidemic histograms aid in containing an epidemic of PSP.

Looking at the place where disease strikes may also provide useful clues as to how PSP was acquired. While PSP patterns may be evident from a frequency distribution of the addresses of the patients, it is often useful to draw a map. Fig. 1 shows how PSP was distributed in Victorias, Negros Occidental, Philippines. The density of patients represented the route of the ambulant vendors who unsuspectingly sold contaminated mussels.

Looking at the persons who get sick can also be very helpful in learning how PSP was acquired. The most important characteristics to look for are age, sex, occupation, food items eaten and food source. By asking patients the day of purchase and the person from whom they bought the mussels, the investigators of the PSP epidemic in Negros Occidental, Philippines, were able to identify the contaminated mussel farm.

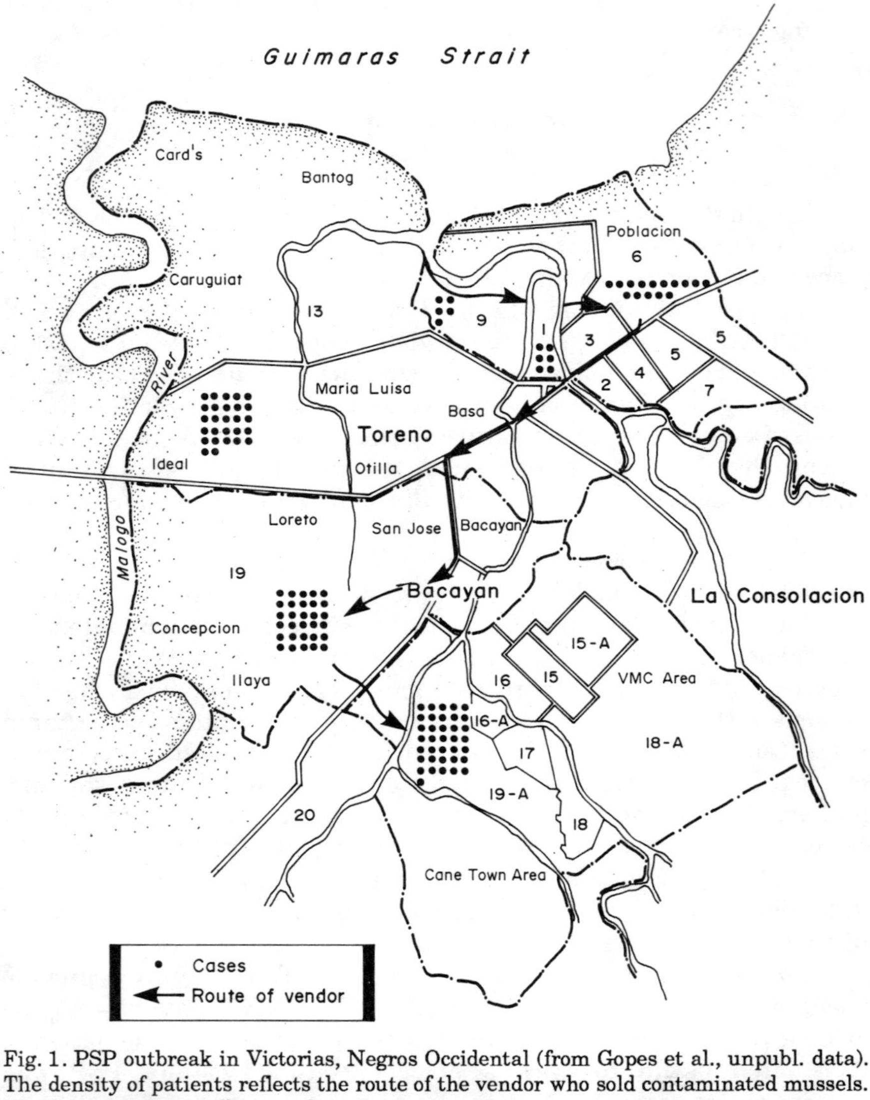

Fig. 1. PSP outbreak in Victorias, Negros Occidental (from Gopes et al., unpubl. data). The density of patients reflects the route of the vendor who sold contaminated mussels. Numbers represent villages or barangays.

Identifying the Cause of the Epidemic

Assemble the dietary data of the patients so that it can be clearly interpreted. Construct a table containing the marine fishery products ingested by the patients and record the number of patients who ate that product. By calculating the percentage of patients who ate a particular marine fishery product, the suspect food item can be identified, which is the food item ingested by the greatest percentage of patients. Transvectors, the food item containing PSP, are identified by both epidemiologic analysis and documenting the PSP toxin in the food item.

Implementing Control Measures

When patients have been counted and transvectors identified, measures to contain the epidemic should be immediately undertaken. Two possible forms are available: (a) health education campaigns and (b) quarantine (Fig. 2). An example of a "Red Tide primer" to be distributed among all people in coastal areas is included in Appendix II. This is the most important part of the investigation, because the whole point of studying epidemics is to keep people from getting sick.

RED TIDE UPDATE

THE RED TIDE UPDATE will appear as a regular weekly item in this paper. Today's report is the 5th of a series of updates to assure that safe and nutritious foods are available to the public.

The Department of Health and the Department of Agriculture, after thorough deliberation and analysis announce that **a small portion of Manila Bay in Bgy. Luz, Limay up to Kapunitan, Orion in Bataan** is being quarantined. Harvesting and transporting of mussels and oysters from these areas are prohibited.

However, tahong and talaba from other portions of Manila Bay, are now safe for human consumption.

Due to the presence of red tide poison, the following products from Bgy. Luz, Limay to Kapunitan, Orion, Bataan; Maqueda Bay and Samar Sea are not yet safe for human consumption.

1. Tahong 2. Talaba 3. Aligi of saltwater crabs in Samar

The warning on the abovestated products still stands to further assure the safety of the public.

Considered safe for human consumption are:

1. **Tahong, talaba, hala-an from Cavite, Parañaque, Bacoor, Navotas, Obando, Bulacan; and Pampanga.**
2. **All cultured and freshwater fish, prawns, shrimps and crabs.**
3. **All fish and seashells not harvested from Maqueda Bay and Samar Sea.**
4. **Saltwater crabs in Manila Bay.**

Meanwhile, the Department of Agriculture will continue to monitor these areas to provide the necessary information to the public.

Fig. 2. Quarantine notice which appeared in Manila newspapers on 15 November 1988.

References

Bres, P. 1986. Public health action in emergencies caused by epidemics. World Health Organization, Geneva. 8 p.

CDC. 1987. Principles of epidemiology: investigation of disease outbreaks. Centers for Disease Control. Atlanta, Georgia. Manual 6.

MacDonald, K.L. and P.M. Griffin. 1982. Food borne disease outbreaks, annual summary. Morbidity and Mortality Weekly Report 1982:35:1-7.

Appendix I

Field Epidemiology Training Program
Department of Health
Philippines

FOOD POISONING INVESTIGATION FORMS
TO DETERMINE THE FOOD THAT CAUSED THE ILLNESS PLEASE MATCH ONE SICK PERSON WITH A HEALTHY PERSON

CASES

NOTE: Sick (cases) persons should be compared to healthy persons to determine if the seafood actually caused his illness.

Date of Interview ____________________
NAME ____________________
BIRTHDATE ____________ SEX ________
ADDRESS COMPLETED ____________________
____
____

DATE OF OFFENDING MEAL ____________________
TIME OF OFFENDING MEAL ____________________

DATE ILL ________ TIME ILL ____________

MEAL CONSISTED OF

	QTY.	NO. TIMES/ WEEK
1. Mussels	______	______
2. Rice	______	______
3. Water	______	______
4. Oysters	______	______
5. Tulya	______	______
6. Hala-an	______	______
7. Fish	______	______
8. Shrimps	______	______
9. Alimasag	______	______
10. Alimango	______	______
11. Squid	______	______
12. Salted fish	______	______
13. Hard drink	______	______
14. Soft drink	______	______
15. Dried fish	______	______
16. Pork	______	______
17. Beef	______	______
18. Chicken	______	______
19. Vegetables	______	______

FOOD HANDLING PROCEDURES:
(This is very important!)

DATE WHEN FOOD WAS BOUGHT ____________________
TIME FOOD WAS BOUGHT ____________________
WHERE WAS FOOD BOUGHT ____________________
FROM WHOM ____________________
WHO COOKED FOOD ____________________

CONTROLS

NOTE: The healthy (controls) person may be a relative or neighbor who should be the same age as the sick person

Date of Interview of Interview ____________________
NAME ____________________
BIRTHDATE ____________ SEX ________
ADDRESS COMPLETED ____________________
____
____

DATE OF MEAL SAME AS CASE ____________________
TIME OF MEAL SAME AS CASE ____________________

MEAL CONSISTED OF

	QTY.	NO. TIMES/ WEEK
1. Mussel	______	______
2. Rice	______	______
3. Water	______	______
4. Clams	______	______
5. Tulya	______	______
6. Hala-an	______	______
7. Fish	______	______
8. Shrimps	______	______
9. Alimasag	______	______
10. Alimango	______	______
11. Squid	______	______
12. Salted fish	______	______
13. Hard drink	______	______
14. Soft drink	______	______
15. Dried fish	______	______
16. Pork	______	______
17. Beef	______	______
18. Chicken	______	______
19. Vegetables	______	______

FOOD HANDLING PROCEDURE I S:
(This is very important!)

DATE WHEN FOOD WAS BOUGHT ____________________
TIME WHEN FOOD WAS BOUGHT ____________________
WHERE WAS FOOD BOUGHT ____________________
FROM WHOM ____________________
WHO COOKED FOOD ____________________

Continued

Appendix I. Continued

PRE-COOKING PRACTICES:
DID YOU WASH THE FOOD BEFORE COOKING? ______
PLACED FOOD IN FREEZER BEFORE COOKING? ______

HOW DID YOU COOK FOOD? ______
UTENSIL CLEANSING AND USE:
DID YOU WAS THE
COOKING UTENSILS ______
SERVING UTENSILS ______

DID YOU USE THE FF: (PLS. CHECK)
______ SPOON
______ HANDS

WERE YOU SICK BEFORE THIS MEAL? ______

PRE-COOKING PRACTICES:
DID YOU WASH THE FOOD BEFORE COOKING? ______
PLACED FOOD IN FREEZER BEFORE COOKING? ______

HOW DID YOU COOK FOOD? ______
UTENSIL CLEANSING AND USE:
DID YOU WAS THE
COOKING UTENSILS ______
SERVING UTENSILS ______

DID YOU USE THE FF: (PLS. CHECK)
______ SPOON
______ HANDS

WERE YOU SICK BEFORE THIS MEAL? ______

THE FOLLOWING QUESTIONS ARE FOR SICK (CASES) PEOPLE ONLY

COMPLAINTS (Check if present):

______ NUMBNESS OF LIPS
______ NUMBNESS OF HANDS/ARMS
______ NUMBNESS OF FEET
______ PARESTHESIA
______ DIARRHEA
______ DYSPNEA
______ HINDI MAKAKAGALAW
______ GEN. BODY MALAISE
______ DOUBLE VISION
______ HEADACHE
______ FELT HOT
______ DIZZY
______ DYSTHESIA

______ DYSPHAGIA
______ NUMBNESS OF HANDS/ARMS
______ VOMITING
______ INABILITY TO AMBULATE
______ PARESIS
______ LIGHT HEADED SENSATION
______ PRURITUS
______ DYSPHONIA
______ CYANOTIC NAILBEDS
______ CYANOTIC LIPS
______ HIVES
______ NAUSEA
______ SHORT TONGUE SENSATION

MEDICAL HISTORY:

SEEN BY PHYSICIAN ______ (YES, NO)
HOSPITALIZED ______ (YES, NO)

DATE SEEN/HOSPITALIZED ______
TIME SEEN/HOSPITALIZED ______

DATE WELL ______
TIME WELL ______

DEAD ______ (YES, NO)
CAUSE OF DEATH ______

Appendix II. Example of "Red Tide Primer"

Red Tide refers to the reddish-brown discoloration of seawater caused by the proliferation of microscopic organisms called dinoflagellates. The species causing red tide in the Philippines is named *Pyrodinium bahamense* var. *compressum.* It thrives in coastal waters and lagoons under conditions of high salinity and does not survive in freshwater fishponds. Toxic red tides are not unique to the Philippines but are known from all over the world, from Australia, Brunei, Canada, Europe, India, Indonesia, Japan, North America, Papua New Guinea, Singapore and Thailand.

Humans can become ill from eating seafood products contaminated with red tide organisms. The most notorious seafood causing illnesses are bivalve shellfish such as the *green mussel, oysters, scallops, cockles* and *limpets* but the gills and guts of small fish such as *sardines* and *anchovies* can also become contaminated. Large fish are safe even if caught from water heavily contaminated with red tide. Fish, shrimps, prawns, crabs and other products grown in ponds or freshwater are safe because red tide organisms do not survive under conditions of low salinity.

The illness following consumption of green mussels is called *paralytic shellfish poisoning* (PSP). Diagnostic symptoms are tingling or burning sensation of the lips, gums, tongue and face, progressing to the neck, arms, fingertips and toes. In severe cases, inability to walk, difficulty in breathing, swallowing and speaking do occur and patients may die from respiratory paralysis. Shellfish poisoning is caused by the neurotoxins from the dinoflagellate which are concentrated by feeding shellfish and toxins can also be contained in dinoflagellates held in small fishes' stomachs. The toxins are stable to heat and therefore are not destroyed by cooking; they are soluble in water and can be present in mussel broth; using vinegar in the cooking of mussels may increase the toxicity to humans. The amount of toxin present in mussels varies dependent upon the intensity of the red tide. Any sample containing more than 80 μg of toxin per 100 g of mussel meal is considered dangerous for human consumption.

There is no fool-proof antidote. A useful household remedy consists of drinking one table spoon of *baking powder* (sodium bicarbonate) dissolved in one glass of water or drinking *coconut milk with brown sugar.*